Civil Engineering Heritage
Ireland

R.C.Cox, MA, MAI, PhD, CEng, FIEI, MICE
M.H.Gould, BSc, PhD, CEng, MICE

Other books in the Civil Engineering Heritage series:
Northern England. R. W. Rennison
Southern England. R. A. Otter
Eastern and Central England. E. A. Labrum
Wales and West Central England. R. Cragg

Future titles in the series:
Scotland
London and the Thames Valley

Ireland
Published for the Institution of Civil Engineers and the Institution of Engineers of Ireland by Thomas Telford Publications, 1 Heron Quay, London E14 4JD

First published 1998

A CIP record exists for this book
ISBN 0 7277 2627 7

Typeset by Katy Carter
Printed and bound by Interprint, Malta

Preface

A civil engineer may be described as a person having responsibility for the design, construction and maintenance of a nation's infrastructure. Thus civil engineers have been at the forefront in the provision of transport systems, whether they be canals, railways, and roads, or the harbours and airports associated with sea and air travel. They play a dominant role in the provision of water supply and sewerage systems, which remain the principal defence against a range of epidemic diseases. Other works commonly undertaken by civil engineers include the construction of power stations and offshore structures for the oil and gas industries. Civil engineers are also involved in water resource management, arterial drainage, coastal protection, and the construction of a wide range of industrial and domestic buildings.

In 1971 the Institution of Civil Engineers (ICE) set up the Panel for Historical Engineering Works (PHEW). The initial objective of the panel was to compile records of civil engineering works of interest throughout Great Britain and Ireland. The works covered by these records were selected for their technical interest, innovation, durability or visual attraction. The records, which are subject to revision in the light of new evidence, form an archive, which has become the principal repository for records of civil engineering works and is regarded as a leading source of information on such works. Although the modern profession of civil engineering can be reasonably dated from the foundation of the ICE in 1818, the archive contains records of many older works, some of great antiquity.

The Institution of Civil Engineers of Ireland, now the Institution of Engineers of Ireland (IEI), was founded in 1835. In 1983 the IEI established a Heritage Society, and in 1987 the society was charged with the task of compiling records of historical engineering works of interest within the Republic of Ireland. This work was undertaken by Ron Cox, with the aid of grants from the National Heritage Council of Ireland and the support of the Department of Civil, Structural and Environmental Engineering at Trinity College Dublin. A Centre for Civil Engineering Heritage was established at Trinity College Dublin in 1995.

The Historical Engineering Committee of the ICE (Northern Ireland Association) has fulfilled a similar role in relation to historical civil engineering works within Northern Ireland. This work has, in recent times, been directed by Michael Gould. Close contact is maintained between the PHEW and the IEI Heritage Society in relation to the work of recording and the maintenance of the records in the archives of both the ICE in London and the IEI in Dublin.

The series of volumes entitled *Civil Engineering Heritage* seeks to make available to a wider public, information contained in these archives, together with research undertaken by the members of the PHEW and the IEI Heritage Society. The vast majority of the archival records consulted for the present volume *Civil Engineering Heritage: Ireland* has been compiled over recent years by one or other of the joint authors.

The authors would like to thank the following for their assistance in this task: Belfast City Council, City and Borough Engineers throughout Ireland, Commissioners of Irish Lights, County and Urban District Engineers, the Department of the Environment, the Department of the Environment for Northern Ireland, the Historical Engineering Committee of the Institution of Civil Engineers (NI Association), Iarnród Éireann, the Inland Waterways Association of Ireland, the Institution of Engineers of Ireland Heritage Society, the Irish Railway Record Society, the National Heritage Council, the National Library of Ireland, Northern Ireland Railways, the Office of Public Works, Trinity College Dublin Library, the Ordnance Survey of Ireland and the Ordnance Survey of Northern Ireland.

Many individuals have assisted with the research for this volume and the authors would especially like to thank the following: Alan Cooper, Brendan O'Donoghue, Dermot O'Dwyer, Don McQuillan, Gerry Daly, Jackson McCormick, Jock McEvoy, John Callanan, John de Courcy, Lisa Cox, Mary Murphy, Michael Barry, Michael Costeloe, Michael Lynch, Michael Taylor, Mike Chrimes, Paul Duffy, Peter O'Keeffe, Ruth Heard, and Simon Perry.

Last, but by no means least, a special 'thank you' to Bryan O'Loughlin, a past Vice-Chairman and former Technical Secretary of PHEW, for checking the draft of the text and for his invaluable advice and encouragement during the writing of this book.

It is hoped that both those with a civil engineering background, and the more general reader, will be assisted by this volume in better understanding the contribution made by the civil engineering profession in respect of the development of Ireland's infrastructure, and that they will gain a greater insight into the diversity of expertise and experience which has contributed to the civil engineering heritage of Ireland.

Ron Cox and Michael Gould

Contents

Front cover: Grand Canal, Dublin (R. C. Cox)

Title page: Fastnet Rock Lighthouse (Commissioners of Irish Lights)

Metric equivalents

Imperial measurements have generally been adopted to give the dimensions of the works described, as this system was used in the design of the great majority of them. Where modern structures have been designed to the metric system, these units have been used in the text.

The following are the metric equivalents of the Imperial units used.

Length	1 inch = 25.4 millimetres
	1 foot = 0.3048 metre
	1 yard = 0.9144 metre
	1 mile = 1.609 kilometres
Area	1 square inch = 645.2 square millimetres
	1 square foot = 0.0929 square metre
	1 acre = 0.4047 hectare
	1 square mile = 259 hectares
Volume	1 gallon = 4.546 litres
	1 million gallons = 4546 cubic metres
	1 cubic yard = 0.7646 cubic metre
Mass	1 pound = 0.4536 kilogram
	1 Imperial ton = 1.016 tonnes
Power	1 horse power (h.p.) = 0.7457 kilowatt
Pressure	1 pound force per square inch = 0.06895 bar

N

Belfast City
and District
④

Donegal
Londonderry
Antrim

⑤
Ulster
Northern Ireland
Tyrone

Leitrim Fermanagh
Armagh Down

Sligo
Monaghan

Mayo
Cavan

⑥
Connaught
Roscommon
Louth

Longford
② Meath
North Leinster Dublin

Westmeath
① Dublin City and District

Galway
Offaly
Kildare

③
Wicklow
South Leinster

Clare
Laois

Republic
of
Ireland
Carlow

Kilkenny
Wexford

Limerick
Tipperary

⑦
Waterford
Munster

Kerry

Cork

0   Scale in km   80

0   Scale in miles   50

Figure 1

Introduction

This volume covers the whole of the island of Ireland. As with the other volumes in the series, each chapter relates to a defined geographical area. A location map, a list of the described sites and a brief introduction are provided for each chapter. The Historical Engineering Work (HEW) number, under which the work is registered with the Institution of Civil Engineers, is also given for each site.

References for individual items are included, where appropriate, and there is appended a select bibliography relating to civil engineering heritage in Ireland. The Ordnance Survey Irish national grid reference is given for each site.

The items have been selected in order to illustrate some aspect of the historic development of civil engineering skills or the scope of activity undertaken by the civil engineering profession in Ireland.

The Island of Ireland

Ireland is divided historically into four provinces: Ulster (nine counties) in the north, Munster (six counties) in the south, Leinster (twelve counties) in the east and midlands, and Connaught (five counties) in the west. These divisions have been used in this present work, but the large province of Leinster has been further subdivided into two regions, north and south. Dublin and Belfast have also been treated separately.

Ireland has a land area of 32 544 sq. miles. Its greatest length is 302 miles, its greatest width 189 miles, and its coastline extends for over 2000 miles. The coastline is dotted with many small harbours and piers. Many of these were built in the nineteenth century with government grants in support of coastal trading and an expanding fishing industry. Each presented its own civil engineering problems and would be worthy of a separate study.

The country consists of an undulating central plain of limestone, almost completely encircled by a coastal belt of highlands of varying geological structure. The central plain is extensively covered with peat bogs and glacial deposits of sand and clay; numerous lakes dot its surface. The River Shannon, which drains this area, has a basin equal to more than 20 per cent of the area of the entire island.

While much of the peat cover has been removed for fuel (including its use for the generation of electricity), the central Bog of Allen and other extensive areas of deep bog still form significant barriers to good communication. Eighteenth-century engineers overseeing the construction of the Grand Canal had to formulate techniques aimed at overcoming the inherent instability of these bogs.

The rivers of Ireland are generally small, although many discharge into deep, often ice-formed, estuaries. The River Shannon is a notable exception. Whilst its north–south course provided a useful waterway, it formed a definite barrier to east–west communications. Most river systems have over the years been subjected, to a greater or lesser extent, to arterial drainage schemes, and many old stone bridges now have concrete anti-scour skirts added to their piers and abutments.

It has been estimated that, in Tudor times, the populations of Ireland and England were comparable. Much small-scale iron smelting was undertaken, using charcoal obtained from the indigenous oak woods. However, due to successive glaciations, Ireland lacks the wealth of minerals found in Britain, in particular significant supplies of good quality coal. Thus, while Britain forged ahead into the Industrial Revolution, Ireland, having exhausted its oak woods, stood still in terms of industrial development. The population was to increase until about 1830, but a series of famines, most notably the Great Famine period of 1845–48, led to many deaths and a significant increase in emigration. The land to the west is generally of poor quality and is often subject to adverse weather conditions emanating from the Atlantic Ocean. In 1882 certain of these areas, those with a very low per capita valuation, were classed as 'congested' and were assisted by special aid. From the middle of the last century, the overall population started to fall, a decline only reversed in recent years. The present population of the island is around 4.5 million.

Ireland's infrastructure has not only been subject to the same periods of unrest as the rest of Britain, but in addition, suffered the destructive effects of the Elizabethan and Williamite wars. Each of these episodes was followed by a series of Plantations, with land being given to settlers from England and Scotland. Many of the settlers brought with them their own construction techniques and expertise. One result of this periodic warfare is that there are few works in Ireland dating from before about 1700.

Records have also suffered and are generally sparse. Even in this century much was lost with the burning of the Public Records Office in

Dublin in 1922 during the Civil War. In particular, there is a scarcity of Grand Jury records, which gave details of the construction of roads and bridges. An air raid on Belfast in 1940 destroyed the drawing office of the Northern Counties Committee of the London, Midland and Scottish Railway, a company pioneering the use of reinforced and precast concrete for railway works.

Central Administration

For a period in the eighteenth century, Ireland had its own parliament, based in Dublin. The Irish Parliament was very supportive of certain industries, such as linen, and of certain civil engineering works, especially canals. Following the Act of Union in 1800, power was transferred back to London.

The Westminster Parliament is sometimes considered to have been less favourable to Ireland, but the establishment in 1832 of a Public Works Loan Board, able to give grants and/or loans, was to provide an important spur to the development of Ireland's infrastructure. Additionally, government support of public works was often seen as a way of assisting the unemployed poor.

From about 1880, Ireland was subjected to periods of agitation, which resulted in 1921 in the formation of the Irish Free State (Republic of Ireland from 1948). The new state covered 26 of the 32 counties, the six forming Northern Ireland remaining within the United Kingdom. The formation of the Free State was not acceptable to all and a period of civil war followed, during which many civil engineering works were attacked. Northern Ireland has been subject since the mid-1950s to further periods of unrest, again resulting in occasional attacks on civil engineering structures.

In this volume, the term 'Ireland' refers to the whole of the island. Where a distinction is required, 'Republic of Ireland' and 'Northern Ireland' are used to signify the respective jurisdictions.

Local Administration

Local government administration was slow in coming to rural Ireland. From 1710 the Grand Juries were enabled to raise funds by Presentment for a limited range of works – most notably, from the civil engineering point of view, for roads and bridges. Later, guarantees were also given to support the operation of railways. Until the advent of public works loans, moneys for these works had to be raised from the larger landowners and progress was slow. Each Grand Jury had to decide its own policy in such

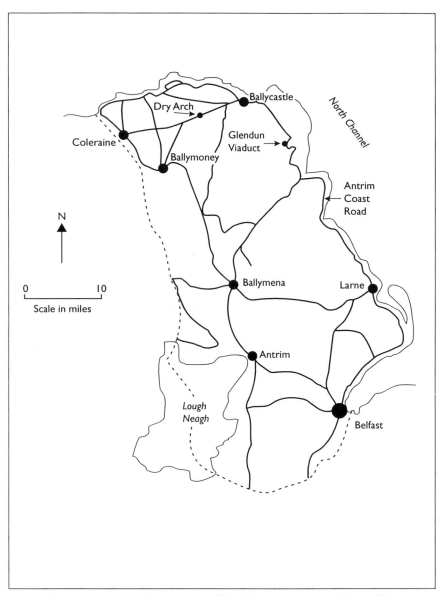

Figure 2. Grant-aided roads in County Antrim, 1832–1899 (Annual Reports of Board of Works, Ireland)

matters, but pressure for road improvement also came from the Irish Post Office (who had the unenviable task of running mail coaches to time), and most counties in Ireland undertook major road and bridge improvements in the nineteenth century.

County surveyorships were established in 1834, and in 1899 the civil engineering functions of the Grand Jury were passed to the newly established county councils. These bodies continued after the establishment of the separate jurisdictions in 1921, although in 1974 the county councils in Northern Ireland were abolished, the functions passing to government departments (most notably to the Department of the Environment for Northern Ireland).

The map in Figure 2 shows, by way of an example, the roads reconstructed or realigned in County Antrim, and supported by public works loans or grants made to the Grand Jury, during the period 1832 to 1899.

Progress in the provision of sanitary services was equally slow. It was not until the second Public Health (Ireland) Act of 1878, which required Boards of Guardians for the poor to take on the role of rural sanitary authorities and provide the necessary funds, that any major progress was achieved. In 1899 these duties passed to new rural and urban district councils. In 1926, RDCs were abolished in the Republic of Ireland, but both continued in Northern Ireland until the 1974 reorganization. The replacement district councils in Northern Ireland have few civil engineering functions.

Inland Navigation

Although Ireland lacked major mineral resources, the hope that an indigenous coal industry would be a practical proposition led to the construction of the Newry Canal, the first modern summit level canal in either Britain or Ireland. Other lines of canal followed, including the east–west lines of the Grand (with a branch southwards to the Barrow Navigation) and Royal Canals centred on Dublin. In addition, the Shannon, Lower Bann and Lagan Navigations, like the Barrow Navigation, followed the lines of existing rivers. The last major line of canal to be constructed, the Ulster Canal from Lough Neagh to Lough Erne, was built to an unusually small width, thus precluding any possibility of an integrated network. Works were generally easy without the many tunnels and aqueducts found in Britain. Although some canals enjoyed a modest period of success, the lack of significant mineral traffic soon led to their falling into disuse, especially where there was competition from the railways. The introduction of pleasure craft on canals and navigations has been a gradual process, but they are now firmly established as an important recreational and tourist amenity. Much canal restoration work has been put in hand in recent times as a result.

Arterial Drainage

In Ireland, arterial drainage generally refers to the deepening, widening and occasional removal of extreme meanders of rivers and the removal of obstructions to the smooth and efficient flow of rivers or streams.

Historically, such arterial drainage can be traced to the early eighteenth century, but it was not until the middle of the last century that drainage work was carried out in a coordinated way on a nationwide basis under the supervision and funding of the Board of Works. Even then, progress was slow, except during the famine years, when drainage schemes were used to give employment to the poor. Subsequent Acts introduced during that century did little to improve the situation.

From an engineering point of view, one of the most significant results of arterial drainage in the nineteenth century was the collection of vast amounts of rainfall and run-off data. This, together with a study of flow in open channels, led Robert Manning, Chief Engineer of the Irish Board of Works between 1874 and 1891, to develop his empirical equation for open channel flow, which is still widely used.

During the present century, there have been a number of Arterial Drainage Acts passed in the Republic of Ireland, notably those of 1927 and 1945. Under the former Act, mechanical excavators were first employed on the Barrow Drainage Scheme. The Brosna Drainage Scheme, the first to result from the latter Act, serves to illustrate the extent of engineering works involved in such schemes. In a catchment area of 312 000 acres, 330 miles of channels were improved. This necessitated alterations to many structures, including the underpinning of over 200 bridges and the construction of nearly 600 new accommodation bridges. In Northern Ireland, arterial drainage was undertaken by the drainage division of the Department of Agriculture (now the Rivers Agency).

Railways

Canal and railway construction in Ireland was, in civil engineering terms, generally relatively easy. There are only a few major structures and even fewer tunnels. In 1859 it was calculated that the average cost of railways in Ireland was only £15 000 per mile compared to £39 000 per mile in England.

The first line of railway to be built in Ireland was between Dublin and Kingstown (now Dun Laoghaire) using the English gauge of 4 ft 8½ in.

By contrast, the second line, the Ulster Railway, used a gauge of 6 ft 2 in. After some lively debate, a compromise Irish gauge of 5 ft 3 in. was adopted throughout the country.

The lack of commercially viable deposits of minerals, particularly steam coal, meant that the Irish rail network was limited, unlike in Britain, where a plethora of lines resulted from the many manufacturing and marketing opportunities presented by the Industrial Revolution. In Ireland, some rural branches were laid to small market towns and low receipts resulted. The success of the 3 ft gauge railway in Glenariff in County Antrim resulted in a number of rural lines being built to this gauge. Only in County Donegal, however, was there any sort of comprehensive 3 ft gauge network. All these lightly used rural lines have now succumbed to road transport. In the Republic of Ireland the main railways lines have been retained, and it is now government policy to retain any closed track *in situ* for ten years. In Northern Ireland, by contrast, only the rump of the former network remains and motorways were constructed in Northern Ireland much earlier than in the Republic. More recently, the Dublin DART rapid transit rail system, the upgrading to high-speed operation of the Belfast–Dublin and Dublin–Cork routes, and the cross-harbour link in Belfast, suggest a secure future for what remains.

Arch Bridges

It has been reckoned that Ireland possesses some 25 000 masonry arch bridges of over 6 ft span, many of them erected in the late eighteenth and early nineteenth centuries with grants from the Grand Juries or other agencies.

The earliest bridges are generally constructed of undressed stone (sometimes called field stone), later laid in courses. As a knowledge of stone working grew, dressed stone came into use, initially in the arch barrels (or voussoirs) and copings, but later in the whole bridge. Ashlar masonry structures also exist in Ireland. The coming of the canals, and later the railways, allowed for the transport of bricks from, for example, Coalisland in County Tyrone. These were used for arch barrels with stone facings or for complete bridges.

Bridges in iron are much less common in Ireland, due to the high cost of imported materials. Although some local foundries existed, most notably J. and R. Mallet, Turner, and Courtney, Stephens and Bailey in Dublin, their output was limited. However, the ability of local foundries, such as that at Armagh, to cast many small parts, led to the use of the Dredge bridge patent in Ulster.

Reinforced concrete was slow to be adopted in Ireland. Most early concrete works appear to have been designed by specialist companies in Britain, although the plans often bore the name of the local county surveyor. Given the relatively smaller number of concrete works, Ireland does possess a surprising number of notable concrete structures, some of which were advanced for their time in terms of design and/or methods of construction.

On lightly trafficked routes, the masonry arch being inherently strong under compression, such bridges will rarely fail unless there has been movement of the piers and/or abutments due, for example, to the scouring action of flood waters. On the other hand, the situation on national primary routes may often give rise for concern due to the increased axle loads now permitted for commercial transport. Repeated heavy loading and severe vibration causes the fill between the spandrel walls to consolidate and expand outwards, causing bulging of the walls and cracking of the arch rings.

In Northern Ireland, the Department of the Environment recently embarked on a wide-ranging bridge assessment and strengthening programme with a view to permitting lorries with heavier axle loads to traverse most of the road network. Unfortunately, in some cases, the need to strengthen is in conflict with any desire to preserve the old bridge structure or its original structural integrity. As a consequence, many brick jack-arch bridges, in particular, have been removed.

In the Republic of Ireland, where the legal responsibility for road bridges lies with the local authorities, the survey and structural assessment of masonry road bridges has been carried out in a more piecemeal fashion in recent years and a number of structures have been rehabilitated and strengthened to cope with increased loading. There is no central database of such bridges, records being maintained by each local authority.

Protective Legislation

Although Northern Ireland represents the smaller part of the island, the protection of buildings of historical and architectural importance by statutory listing is well established. Following the practice in Britain, listing under the planning legislation was introduced in 1969. In certain circumstances, a partial grant is available to assist with maintenance and restoration.

Under Northern Ireland legislation, the definition of 'building' includes any complete structure. Some 300 bridges of all types are listed, as well as other examples of civil engineering structures.

The current system of listing in the Republic of Ireland differs from that in Britain and Northern Ireland in that there is currently no statutory grading of listed buildings. However, there is generally a commitment from successive governments to improve the protection for listed buildings, including placing the system of listed buildings on a statutory basis and introducing incentives for the proper upkeep and maintenance of such buildings. It is to be hoped that the definition of 'building' to be adopted will be similar to that used in Northern Ireland and that many important civil engineering structures will qualify for listing.

In the meantime, the legislation of most relevance to the listing of structures in the Republic continues to be the Planning Acts. Section 19 of the Local Government (Planning and Development) Act 1963 obliges each planning authority to maintain a development plan for its area, setting out certain mandatory development objectives. Local authorities can provide a degree of protection for buildings, etc. in the development plan. Generally, buildings, etc. in 'List 1' are those whose preservation is considered essential, while in 'List 2' are those whose preservation will be considered if and when a planning application is made. The concern is that most, if not all, civil engineering structures in the Republic, if listed, are in 'List 2', which offers only a small degree of protection. A constant watch, therefore, needs to be kept for any proposals to demolish or alter such structures.

Surveys and Maps

The Down Survey was completed by William Petty in 1657 for the purpose of recording lands forfeited following the Cromwellian plantation. Some sheets were published at scales as large as four inches to the mile. The survey covered about two-thirds of the island and recorded bridges and mills. Hermann Moll's map of 1714 shows all the major roads, whilst Taylor and Skinner's strip maps of 1778 contains the names of a number of bridges.

It was not until the establishment of the Ordnance Survey in Ireland in the 1830s that accurate and detailed maps for the whole of the island were prepared and published. The first edition of the Ordnance Survey six inch to the mile map series was accompanied by written Memoirs, prepared by individual surveyors for each civil parish. These memoirs are, however, of variable quality and are incomplete. The Institute of Irish Studies at the Queen's University in Belfast is currently in the process of publishing those covering Ulster.

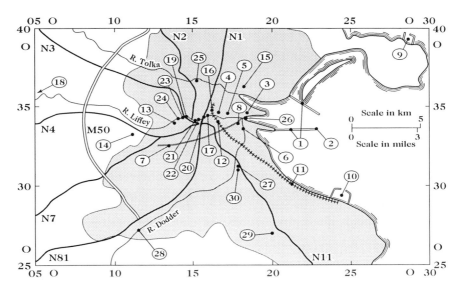

1. The Bull Walls
2. Poolbeg Lighthouse
3. Dublin Port
4. Custom House Docks and Stack 'A' Warehouse
5. Scherzer Lift Bridges, Dublin
6. Ringsend Bridge
7. Grand Canal (Circular Line)
8. Grand Canal Dock
9. Howth Harbour
10. Dun Laoghaire Harbour
11. Dublin and Kingstown Railway
12. Westland Row Train Shed
13. Heuston Station and Train Shed
14. Inchicore Railway Works
15. Fairview Rail Bridge
16. Liffey Viaduct ('Loop' Line)
17. Butt Bridge
18. Lucan Bridge
19. Liam Mellowes Bridge
20. O'Connell Bridge
21. Grattan Bridge
22. Liffey Bridge
23. Rory O'More Bridge
24. Sean Heuston Bridge
25. Glasnevin Curvilinear Glass Houses
26. Liffey Tunnel
27. Donnybrook Bus Garage
28. Balrothery Weir
29. Vartry Aqueduct and Stillorgan Reservoirs
30. UCD Water Tower

1. Dublin City and District

Dublin, which is situated at the head of Dublin Bay, has grown from being a ninth-century Viking trading settlement to an important European cultural and business centre. The city was, for many centuries, the administrative centre for Ireland and, since 1921, the capital of the Irish Free State, which officially became the Republic of Ireland in 1948. As a consequence, the city has seen a steady rise in population over the years. The importance of good communications with London resulted in the construction of Kingstown (now Dun Laoghaire) Harbour. Designed by the Rennies, it was for many years the largest enclosed harbour in the world. John Rennie also designed Howth Harbour, initially the landfall for the mails being carried to Dublin along Telford's London to Holyhead road.

The city lies mostly on a coastal extension of the central limestone basin. This is overlain with glacial deposits, such as boulder clay and estuarine sands and gravels. At the southern extremity rise the granite foothills of the Wicklow Mountains.

The main rivers flowing through the city are the Liffey, Dodder and Tolka. Originally, these flowed out over extensive mud flats, but they have been gradually forced by the construction of sea walls and associated land reclamation schemes, to discharge their waters into the bay in a more regulated way. The location of the River Liffey, lying through the centre of the city, has necessitated the building of many bridges. There are a great variety of designs, many of masonry, but some of iron, whilst others have been constructed of more modern materials.

The port of Dublin has been developed over a long period of time, and this has involved some major engineering works, such as the construction of docks and quay walls. Excellent examples of nineteenth-century structural engineering may be seen in the roofs of Heuston and Connolly stations and in the large glasshouses at the Botanic Gardens at Glasnevin.

In the early nineteenth century, Dublin was connected by canal with the River Shannon and trade developed along the waterways, until the railways offered a much faster and more convenient alternative mode of transportation. The rivers of Dublin provided the means of driving great numbers of mills, and supplied power for a variety of indigenous industries, such as grain milling, tanning, brewing and distilling. Steam power gradually replaced water power, but coal had to be imported, mostly from Wales.

Dubliners were the first in Ireland to have a public railway and the line along the coast to Kingstown, opened in 1834, was a resounding success.

In the 1840s and 1850s, Dublin was connected by rail with the south and west of the country, and large terminal stations were built in the city. The coastal link with Belfast was finally achieved in 1855 with the completion of the Boyne Viaduct and Bridge, but Wexford was not reached from Dublin by rail until 1872.

Dublin had a limited public water supply from as early as 1244. Supplies were also extracted from the canals during the first half of the nineteenth century. To meet the demands of a growing population and the requirements of public health legislation, a scheme to bring treated water to the city from an impounded reservoir in County Wicklow was completed in the 1860s, and similar schemes followed. The first main drainage and sewage disposal scheme for the city was completed in 1906.

With the growth of industrial and commercial activity, the city and surrounding districts have been faced in the 1990s with growing traffic problems. The completion of the motorway system and a port access tunnel will add significantly to the future civil engineering heritage of the Dublin area.

I. The Bull Walls

The city and port of Dublin are situated where the rivers Liffey, Dodder and Tolka discharge into Dublin Bay. Large accumulations of sea sand to the north and south of the river estuaries form what are known respectively as the North and South Bulls. At the beginning of the eighteenth century the rivers flowed out over these strands at low water and formed natural, but constantly shifting, channels.

HEW 3016

O 181 342 to O 232 340 (North Bull wall)

O 211 360 to O 232 344 (South Bull wall)

In 1711, work began on the provision of a straight channel from the city to Ringsend and a jetty extending 7938 ft from Ringsend to the site of the Pigeon House Fort, and thence a distance of 9816 ft to the eastern spit of the South Bull. This jetty was formed of timber caissons. These were built at Ringsend and floated down river to the site, where they were then filled with rubble to sink them to the sea-bed. The jetty was completed in the 1730s and a floating light vessel moored at its extremity. The first section, from Ringsend to the Pigeon House Fort, was replaced in 1748 by a double line of rubble retaining walls with sand infill. The Poolbeg Lighthouse was completed at the end of the timber jetty in 1767. The present South Bull wall was completed in 1796 by the Corporation for Preserving and Improving the Port of Dublin (commonly known as the Ballast Board), who also provided a small harbour at the Pigeon House to shelter vessels from easterly winds.

About 1800, the Board commissioned Captain William Bligh (of *Bounty* fame) and others to survey the harbour and report on what improvements could be made, particularly with regard to increasing the depth of water at the entrance to the port. The final solution decided on was a scheme, proposed by Giles and Halpin in 1819, which involved the construction of a wall or breakwater from the north shore at Clontarf to a point opposite the Poolbeg Lighthouse. Completed in 1825, the North Bull wall is formed of rubble limestone and granite with an opening between the Bull island and the shore spanned by a wooden bridge.

From the end of the bridge, a length of 5500 ft of wall was completed to the full design height, a length of

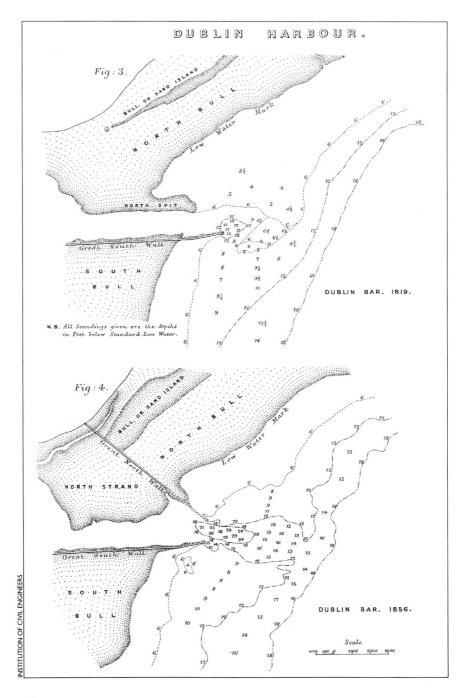

DUBLIN HARBOUR.

Fig: 3.

DUBLIN BAR. 1819.

N.B. All Soundings given are the depths in Feet below Standard Low Water.

Fig: 4.

DUBLIN BAR. 1856.

Scale

1500 ft to high water neap tides, and the remaining section of 500 ft to half-tide level.

On the north side of the bay, during the first half of its ebb, the tide runs westwards towards the bar, and thence southwards in the direction of Dun Laoghaire, while during the last half of its ebb and the whole of its flood, the tide sets eastwards towards the Bailey Lighthouse at the extremity of the Howth peninsula. During the first half of the ebb tide, the tidal and river waters confined within the North and South Bull walls pass partly over the submerged portion of the North Bull wall, and partly through the harbour entrance between the walls. The result is a great increase in the velocity of the current and the removal by scouring action of sand from the bar at a time when the ebb tide is setting eastwards towards deep water. Due to the direction of the currents on the north side of the bay, the flood tide from the south does not normally return the sand to the harbour.

As a result of this remarkable piece of harbour engineering, the depth at low water ordinary spring tides of the navigable channel through the bar had by 1873 increased from around 6 ft to over 16 ft and Sir John Purser Griffith was given to remark that this was 'a noble example of directing the great sources of power in nature for the use and convenience of man'.

GRIFFITH J. P. The improvement of the bar of Dublin Harbour by artificial scour. *Min. Proc. Instn Civ. Engrs*, 1878–79, **58**, 104–145.

GILLIGAN H. A. *A history of the port of Dublin*. Gill and Macmillan, Dublin, 1988, 21–41, 89–96.

Opposite:
Dublin Harbour Bar. (From Min. Proc. Instn Civ. Engrs, 1878–9, **58**, 104–145)

2. Poolbeg Lighthouse

In 1761 the Directors of the Ballast Board raised the necessary finance to continue the building of the Great South or Bull wall. Before completing the eastern section of the wall, it was decided to form a base for a lighthouse at the entrance to Dublin port. Timber caissons were again used and a masonry tower lighthouse was completed in 1767 under the direction of John Smyth, described in the records of the Ballast Board as 'The Architect'.

The Poolbeg Lighthouse, as it has been known to

HEW 3051
O 232 340

generations of Dubliners, replaced a floating light vessel moored at the end of the partially completed wall. Using tallow candles, the light was first exhibited on 29 September 1767, a large crowd watching the event from the end of Sir John Rogerson's Quay. In 1786, it was one of the first in either Britain or Ireland to be converted to use oil for the light, the lamps being replaced in 1888 by an incandescent petroleum vapour burner and in 1964 by electricity supplied by diesel electric generators. In the same year, the lighthouse was made automatic.

In 1813 the tower was increased to its present height of 77 ft 9 in. above high water and the original gallery was removed. The tower is 30 ft 9 in. in diameter at the base, tapering to 19 ft 6 in. at the base of the lantern. The light is at a height of 66 ft above high water.

In the 1870s, in order to provide added protection from winter storms, large rubble concrete blocks, each weighing around one hundred tons, were placed around the base. This was carried out using the floating crane or shears designed by Bindon Blood Stoney and then being used in the building of the extension to the North Wall quays.

At the commencement of the first large-scale surveys of the country by the Ordnance Survey, a national datum for levelling was established at the Poolbeg as the level of low water ordinary spring tides on 8 April 1837. This was found to be 20 ft 11 in. below a permanent mark on the base of the lighthouse tower. The Malin Head datum has now replaced the Poolbeg datum for all modern mapping in Ireland.

GILLIGAN H. A. *A history of the port of Dublin*. Gill and Macmillan, Dublin, 1988, 21–41.

3. Dublin Port

HEWs 3080, 3024 and 3253
O 180 345

The Corporation for Preserving and Improving the Port of Dublin (1786–1867) and the succeeding Dublin Port and Docks Board (established 1867), oversaw the rebuilding of the earlier quay walls along the banks of the river, the construction of a large graving or dry dock, the dredging of the approach channels, and the construction of a number of wet docks.

DUBLIN PORT

The large (No. 1) graving dock (HEW 3024) was built by William Dargan between 1853 and 1860 to accommodate the Holyhead paddle steamers. It is 410 ft long and 69 ft 10 in. wide at the entrance and was, according to a contemporary media account 'one of the most excellent specimens of material and workmanship'. The original dock gates by J. and R. Mallet were replaced in 1931 and the dock has recently been reopened following the fitting of new gates.

North Wall quay extension, Dublin port

The extension of the North Wall quays (HEW 3253) between 1871 and 1885 formed part of what became known as Alexandra Basin. To commence the formation of this basin or wet dock, the port engineer, Bindon Blood Stoney employed large mass-concrete blocks of up to 350 tons in weight, cast on land. They were then placed by a large floating crane on foundations previously prepared by labourers working in a diving bell some 30 ft below the water line. This avoided the need for caissons to be installed around the site.

Purser Griffith was responsible for introducing bucket dredgers (1871) and later suction dredgers (1895) to the port for the purpose of straightening and deepening the approach channels and generally keeping the port operational.

Other achievements in Dublin port, such as the building of the Bull walls (HEW 3016), the Custom House Docks, and the rebuilding of a number of major bridges across the River Liffey, are dealt with elsewhere in this volume.

GILLIGAN H. A. *A history of the port of Dublin*. Gill and Macmillan, Dublin, 1988.

Cox R. C. *Bindon Blood Stoney: biography of a port engineer*. Institution of Engineers of Ireland, Dublin, 1990, 13 ff.

4. Custom House Docks and Stack A Warehouse

HEW 3053 and HEW 3066 O 166 346

A decision by the port authorities towards the end of the eighteenth century to provide deep-water berthage further east of the Custom House resulted in the construction of the Custom House Docks. With a brief from James Gandon, John Rennie completed the Custom House (or Old) Dock in 1796. It was 443 ft by 198 ft and had a tidal lock to the river spanned by a drawbridge. This dock was filled in around 1927. George's Dock (320 ft by 250 ft), with a separate tidal lock entrance, was built further to the east and was opened in 1821. An Inner Dock (556 ft by 279 ft) was completed by 1824. All the docks, together with the earliest warehouses, were designed by Rennie. Later the docks were developed mainly as a warehousing estate.

When the beginnings of the deep-water port were created further down river, the docks fell into disuse as shipping was able to enter the port at all states of the tide, instead of having to enter the wet docks at high tide through a system of locks. However, the warehouses

R. C. COX

continued to be used for storing goods landed in other sections of the port.

Stack A Warehouse was designed by Rennie and erected in the early 1820s as a tobacco warehouse. There are four 154 ft wide bays in a total length of 476 ft. The warehouse was originally 500 ft long but was later foreshortened to permit widening of the quayside roadway. The clear internal height is 20 ft, but a mezzanine floor 154 ft long was added at the north end in 1871.

The roof trusses span 38 ft 6 in. and are supported, in the case of the two outer bays, on 2 ft 3 in. thick masonry walls and, internally, at the quarter points by independent cast-iron arches spanning 18 ft between 9 in. diameter cast-iron columns. These in turn bear on the masonry walls of the vaults underneath.

The trusses are formed from cast-iron compression members of cruciform section and wrought-iron tension members of circular section. Each truss is supported by pairs of struts extending at 45° from the column heads to points on the truss 3 ft from each side. The tops of the inclined struts are tied together by a circular section tie bar, thus completing the triangle of forces. The curved longitudinal beams on top of the columns are 36 in. at the

Interior of Stack A Warehouse prior to restoration

springing, have hollow spandrels, and a rise to span ratio of about 0.3. The iron roof purlins are T-shaped with tapered flanges and webs. It is hoped that the warehouse will be restored and used for some alternative purpose.

Following the return of Irish regiments from the Crimean War in 1856, the government decided to honour them with a banquet. The only building large enough to accommodate the function was the Stack A Warehouse, and it was transformed into a banqueting hall for 3000 guests, the roof trusses being painted red, yellow and blue, the columns blue and the walls white for the occasion.

MEREDITH J. *Around and about the Custom House.* Four Courts Press, Dublin, 1997, 56 ff.

5. Scherzer Lift Bridges, Dublin

HEW 3015
O 173 345

The Royal Canal was completed in 1817 to provide a second line of communication between Dublin and the River Shannon. It joins the River Liffey in Dublin at the North Wall quays to the west of the East Link Bridge and is crossed at this point by twin lifting bridges. These bridges were erected in 1912 by the firm of Spencer & Co. of Melksham in Wiltshire to replace a rolling drawbridge placed there by the Midland Great Western Railway Company in 1860. Sir John Purser Griffith, Chief Engineer to the Dublin Port and Docks Board, based his design on that previously patented in 1893 by William Scherzer of Chicago.

Each bridge consists of two main girders, connected together by 14 floor beams, the floor being composed of patented 'buckled plates' (an invention of the Irish engineer Robert Mallet) resting on and riveted to the floor beams and to joists fitted between the floor beams. To the western ends of the main girders, segmental girders are attached to form the rolling surfaces upon which the bridges bear. The segmental girders are extended so as to carry a large counterweight and the whole structure is suitably braced. To prevent accidental displacement of the bridges on their paths, teeth were formed on the cast steel track plates, and corresponding recesses cut and accurately tooled in the curved track plates of the seg-

R. C. COX

mental girders. Each bridge was worked by electric mo-
tors (since removed), erected on the platform in front of
the counterweight box, but they could be operated manually in case of power failure.

Scherzer Lift
Bridges

A similar pair of bridges was erected by Sir William
Arrol of Glasgow in 1932 over the entrance to the Custom
House Docks further upriver at O 166 346. A 1:40 scale
model of the bridges can be seen in the foyer of the
Museum Building in Trinity College Dublin.

The advantage of the twin bridges was that the stop-
page of quay traffic when vessels needed to enter or leave
the canal system was reduced to a few minutes by alternately raising and lowering each bridge during the locking procedure.

GRIFFITH Sir J. P. The twin Scherzer bridges on the North Wall quay,
Dublin, across the entrance to the Royal Canal and Spencer docks. *Trans.
Instn Civ. Engrs Ir.*, 1912, **38**, 176–204.

6. Ringsend Bridge

The fishing village of Ringsend was established at the
mouth of the River Dodder, which rises in the Wicklow
Mountains and follows a relatively short and precipitous
route to join the River Liffey east of Dublin city centre.

HEW 3047
O 180 337

R.C. COX

Ringsend Bridge

The Dodder is prone to flash flooding, as for example that experienced during Hurricane Charlie in 1986, when much damage was caused.

The old road from Dublin to Ringsend crosses the Dodder at Ringsend by a masonry bridge of an interesting design. The present bridge is a single-span masonry arch bridge constructed using dressed Wicklow granite and replaces an earlier bridge carried away in the flood of 1802.

Built in 1812, the Ringsend Bridge spans 78 ft between abutments and is 27 ft wide between the solid parapets. The rise of the span is 14 ft, and the arch has an elliptical profile. The edges of the arch are chamfered in the manner of the French *cornes de vache*, the chamfer reducing from the springing level to the top of the arch. The abutments are continued downwards as a curve to form a paved river bed. Both these elements of the design ensure good hydraulic flow conditions under the bridge in times of heavy flooding. The spandrel walls are formed by a continuation of the voussoir stones in the arch ring. There is a string course running the full length of the bridge immediately underneath the parapet.

The bridge was designed and built by Dublin Corporation. It is not known who the contractor was, but the quality of the stonework suggests the employment of a master mason.

7. Grand Canal (Circular Line)

The Dublin terminus of the main line of the Grand Canal connecting Dublin with the Shannon Navigation was at James's Street Harbour near the present Guinness Brewery at a point where there is a steep drop down to the River Liffey. William Chapman surveyed a route for a canal to link the main line of the Grand Canal from the First Lock at Suir Road to the river near the mouth of the River Dodder at Ringsend. This plan was commented on by William Jessop and supported by John Macartney and Richard Griffith. The canal (known as the Circular Line) was constructed between 1790 and 1796 under the direction of William Rhodes, James Oates and (for a time) Archibald Millar, all engineers to the Grand Canal Company. The canal cost nearly £57 000, almost five times the original estimate.

HEW 3228
O 128 332 to
O 174 336

The canal is about 3½ miles long and is level from Suir Road as far as Portobello, where a harbour and an hotel were built in 1805. Portobello became the terminus for the passage-boats plying the canal between Dublin and Limerick. A harbour, named after Joseph Huband, was also provided at Dolphin's Barn, but, like those at Portobello and James's Street, has since been filled in. From Portobello the canal then falls by a series of seven single locks to a large dock built by the canal company near Ringsend.

A domestic water supply for Dublin city was for many years taken from the canal at the Eighth Lock at Portobello until superseded by the Vartry river scheme at Roundwood in the 1860s. Commercial consumers, however, notably Guinness, continued to use canal water for many purposes, such as washing and cooling, before switching to mains supplies in the 1980s.

In 1963, Dublin Corporation announced plans to place a main drainage sewer in the bed of the canal and to construct a roadway over it. In 1969, however, the plans were altered and the main interceptor sewer was constructed in a tunnel under the canal, thus preserving the navigation and considerable amenity value of the canal.

DELANY R. *The Grand Canal of Ireland.* David and Charles, Newton Abbot, 1973, 35 ff.

8. Grand Canal Dock

HEW 3082
O 178 339

The Circular Line of the Grand Canal commences at the First Lock on the main line and extends for about $3\frac{1}{2}$ miles in a circular arc through the south city to join the River Liffey at a dock constructed near Ringsend. The dock was constructed between 1792 and 1796, the work being supervised by Edward Chapman, a son of William Chapman, who acted as consultant to the project along with William Jessop. Jessop had earlier drawn up plans for the dock and designed the entrance gates, but left much of the implementation of the project to Chapman. The contractor was John Macartney, who was knighted by the Lord Lieutenant at the official opening of the dock on 23 April 1796.

A river wall, approximately 3000 ft long, had been constructed between 1717 and 1727 from the end of the South quays to near the mouth of the River Dodder (now known as Sir John Rogerson's Quay). By 1760 a bank had also been built along the line of the present South Lotts Road. The area within the banks was reclaimed, together with adjoining areas of the Dodder estuary.

Within this area the Grand Canal Company built a large L-shaped wet dock, divided by the main road to Ringsend into two areas of $16\frac{1}{2}$ and 8 acres. This provided 5300 ft of wharfage. Part of the dock to the west was later filled in after the railway was opened in 1834. The dock was never a commercial success, because of the small size of the locks, siltation problems in the river, and the decision by the Ballast Board to build new docks near the Custom House upstream and on the opposite bank of the River Liffey.

There are three parallel tidal locks of varying capacity: Camden Lock (now derelict), Buckingham Lock (recently restored), and Westmoreland Lock (also now derelict). The canal company also built three graving docks on the land between the docks and the Dodder estuary, the largest being filled in as early as 1851. At this time the Dublin Dockyard Company took a lease on the area and continued in business until 1881. The Ringsend Dockyard Company operated in and around the two smaller graving docks from 1913 until 1963, when they too were filled in.

In 1992 the Office of Public Works built an Inland Waterways Visitors Centre on piles in the inner section of the dock. There are plans for a major redevelopment of the entire area, including the dock.

DELANY R. *The Grand Canal of Ireland.* David and Charles, Newton Abbot, 1973, 56 ff.

HADFIELD C. and SKEMPTON A. W. *William Jessop, Engineer.* David and Charles, Newton Abbot, 1979, 86–98.

9. Howth Harbour

An insufficient depth of water at Pigeon House Harbour and a general shortage of berths in Dublin port were causing such delays to the mails from London that, from 1800 onwards, varying proposals were being put forward by Thomas Rogers, Sir Thomas Hyde Page, and John Rennie Senior for an alternative harbour to be constructed at Howth on the north side of the peninsula to the east of Dublin. One proposal included a canal to connect the harbour with the city.

HEW 3055
O 286 395

In 1807, under the terms of an Act of Parliament (45 Geo.III c.55), a start was made on construction of the east pier at Howth to a plan drawn up by Captain George Taylor, who initially supervised the work. In 1809, about 240 ft of the end of the pier collapsed and Taylor resigned, being succeeded as superintendent of the work by John Aird, acting under the direction of John Rennie.

Rennie had advocated Dun Laoghaire as a more suitable site for an asylum harbour, but work had already started at Howth. Local stone from a quarry at Kilrock was originally used, but large quantities of granite for the inner facing of the piers was later ferried across the bay from Dalkey. Following a recommendation in 1810 by Rennie, a west pier was added and the harbour was substantially completed by 1813. It was not formally established as the mail packet station until 1818, when a

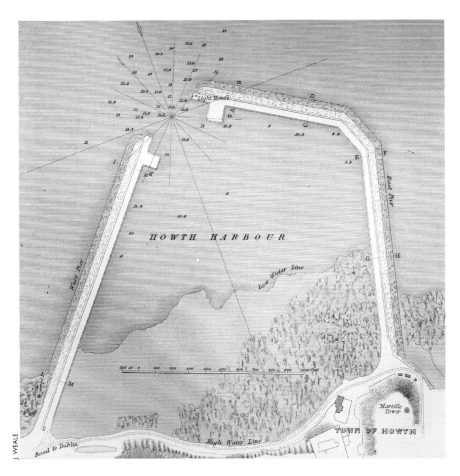

HOWTH HARBOUR

John Rennie's design for Howth Harbour

lighthouse was added at the end of the east pier, together with a light keeper's dwelling (converted to two storeys in 1856).

By 1812, Rennie was able to use a diving bell to assist in the completion of the pier heads, but prior to this, pier foundations below water were prepared using Runcorn stone, the soft texture of the stone making it easy to cut into blocks, each 4 tons in weight. These blocks, when submerged in water, acquired a hard texture and were laid sideways on in front of each other at an angle of 40°. On top of these, horizontal header and stretcher courses were laid in mortar.

The harbour as originally built was formed by two

piers, each 200 ft wide at the base and 85 ft wide at the high water mark, with a 42 ft wide roadway on top. The east pier extended 1300 ft in a NNE direction, then 230 ft NNW and finally 920 ft NW by W. The west pier, commencing approximately 2000 ft along the shoreline, is 1900 ft in length. The harbour entrance is 350 ft wide and the area enclosed is some 52 acres. 100 ft long moles have since been added at the extremities of the piers to counteract wave action.

The harbour was a failure as far as the mail packet ships were concerned and the mails were transferred to Kingstown (Dun Laoghaire) in 1826. Howth Harbour became instead a major fishing port and yachting centre.

John Rennie
(1761–1821)

27

GILLIGAN H. A. *A history of the port of Dublin*. Gill and Macmillan, Dublin, 1988, 109–110.

RENNIE Sir J. *The theory, formation and construction of British and foreign harbours*. John Weale, London, 1854, 199–200.

10. Dun Laoghaire Harbour

HEW 3014
O 245 294

In 1815, eight harbour commissioners were appointed for the purpose of building a new harbour, eastward of the old fishing port of Dunleary, to replace an earlier pier. This had rounded pier heads and a parapet, and was completed in 1767. The 'Dunleary Asylum Harbour' was intended primarily to provide a safe refuge for ships unable to reach Dublin during heavy winter gales.

The initial design of the harbour consisted of a single pier to be carried out about 2800 ft from the shoreline, but in 1817, John Rennie was consulted and subsequently proposed two embracing piers, which later became known as the East and West Piers. In the same year the foundation stone was laid by the Lord Lieutenant of Ireland, Earl Whitworth. Government approval had to be won in stages. Between 1817 and April 1820, 2285 ft of the East Pier was completed. Three years later it had reached 3350 ft, its final length being 4231 ft. The West Pier, begun in 1820, was already 1600 ft long by 1823, and by December 1827 had reached 4140 ft; its final length was 5077 ft.

The base of each pier is 310 ft wide and constructed with blocks of Runcorn sandstone, each 50 cu. ft in volume. From 6 ft below low water and upwards, granite was used. At the top, the pier is 52 ft wide, with a 40 ft promenade on the inner side and an 8 ft to 9 ft parapet wall to protect the upper promenade from waves. The piers enclose an area of about 250 acres.

The rock for forming the piers was quarried at nearby Dalkey and transported to the harbour by means of a funicular railway using six trucks connected together by a continuous chain, each truck carrying 25 tons of granite. The core of the piers consists of granite rubble loosely tipped and allowed to consolidate using the action of the waves (unlike at Donaghadee in County Down, where the rubble masonry was laid in courses).

Following the death of John Rennie in 1821, his son John (later Sir John) Rennie, took over as consultant to the

DUBLIN PORT

project. He proposed two short projecting arms from the ends of the piers, leaving a narrow entrance of about 450 ft. William Cubitt was also consulted and he suggested a much wider entrance. In the event, neither proposal was pursued.

Dun Laoghaire
Harbour

A decision on the form and length of the pier heads was delayed until the Board of Works took over responsibility for the harbour in 1833, the final outcome being to form an entrance to the harbour of about 760 ft in width, with rounded pier heads.

The present East Pier Lighthouse and Battery Fort were completed in 1860. One of the world's first anemometers for recording the strength of the wind, designed by Professor Robinson of Trinity College Dublin, was mounted on the East Pier in 1852.

Mail packet steamers, on the service to Holyhead, were transferred from Howth in 1826 and berthed at a wharf near the present bandstand on the East Pier. A 500 ft wharf was completed in 1837, Traders Wharf in 1855 and the Carlisle Pier with direct rail connection in 1859. A car-ferry terminal was built in 1970 and the rail link finally abandoned in 1981.

Responsibility for the harbour was transferred in 1989

from the Office of Public Works to the Department of the Marine, who are advised by the Dun Laoghaire Harbour Board. A major extension to the vehicle and passenger handling facilities was completed by ASCON Ltd in 1995, including a berth for a new high-speed catamaran passenger/vehicle ferry. This ferry was the largest of its type in the world at the time of its introduction by Stena Line in 1996 on the Holyhead–Dun Laoghaire service.

PEARSON P. *Dun Laoghaire–Kingstown*. O'Brien Press, Dublin, 1981, 13–40.

RENNIE Sir J. *The theory, formation and construction of British and foreign harbours*. John Weale, London, 1854, 197–199.

11. Dublin and Kingstown Railway

HEW 3025
O 166 340 to
O 243 288

An Act of Parliament in September 1831 gave royal assent to a scheme prepared by Alexander Nimmo '... for making and maintaining a railroad from Westland Row in the City of Dublin, to the head of the western pier of the Royal Harbour at Kingstown in the County of Dublin with branches to communicate therewith'.

Nimmo died in January of the following year and, following consultations with Thomas Telford, George Stephenson, Joseph Locke and John Killaly, Charles Blacker Vignoles was appointed in April 1832 by the Board of Works to report on the scheme. He subsequently became engineer to the Dublin and Kingstown Railway Company (D&KR).

This, the first public railway line in Ireland, is double track and was laid initially to the English gauge of 4 ft 8½ in. This was converted to the Irish gauge of 5 ft 3 in. when the line was leased in 1854 by the Dublin and Wicklow Railway Company, which extended the railway to Bray.

From the original terminal station at Westland Row (now Pearse station), the line runs on an embankment between retaining walls as far as Serpentine Avenue in Ballsbridge and thence to Merrion on a low earth embankment. From Merrion to Blackrock and from Seapoint to Dunleary (renamed Kingstown in 1828 to mark the visit of William IV), the line is carried on embankments laid on the strand. These sea embankments comprise two

MERRION TO BOOTERSTOWN.

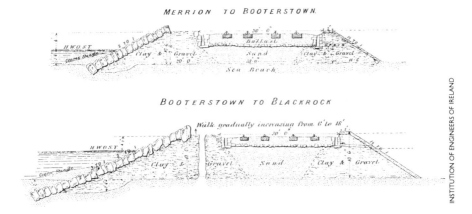

BOOTERSTOWN TO BLACKROCK

INSTITUTION OF ENGINEERS OF IRELAND

parallel bunds consisting of clay / gravel filled in between with sea sand, topped with layers of gravel, and followed by the track ballast. The slopes are pitched with granite blocks and topped with heavy parapet walls.

Dublin and Kingstown Railway: sections through sea embankment

Between Blackrock and Seapoint, the company had to pay substantial compensation to the local landowners who also demanded fishing and bathing lodges, a camera obscura tower, a small harbour with a boat slip, and an iron latticed footbridge leading from their properties across the railway to the sea, all the building work being 'executed in the very best style of Italianate architecture'.

The initial length of the line was about 5½ miles, but it was extended in 1837 a further ½ mile to the mail steamer wharf in the Royal Harbour. This cut off access to the old harbour at Dunleary and a compensation harbour was built (now known as the Coal Harbour). The embankment, boat slips and harbour arrangements were designed by William Cubitt and constructed, as was the entire project, by William Dargan. The present station buildings at Dun Laoghaire were designed by the architect John Skipton Mulvany, the walls of the temporary train shed erected in 1845 being seen today to the north-east of the present platforms.

In 1884, parts of the original train shed at Westland Row were incorporated into the rebuilt terminus, and provision was made to allow for subsequent through working of trains across the River Liffey to Amiens Street (now Connolly Station) via the Liffey Viaduct or Loop Line.

A single-track extension to Dalkey, built by Dargan in 1844, was worked on the atmospheric principle developed by Samuda and Clegg and acted as a feeder to the D&KR until April 1854. The station layout at Dun Laoghaire was altered in 1856 to permit through working to Bray and was further modified in 1985 by the addition of a platform on the south side as part of the electrification of the line, being the first phase of the Dublin Area Rapid Transit System or DART.

The D&KR was one of the world's first and most successful commuter routes. It relied on passengers and mails, gaining little freight traffic because of the development of Dublin as a major port.

GRIERSON T. B. The enlargement of Westland Row terminus, with a sketch of the early history of the Dublin and Kingstown Railway. *Trans. Instn Civ. Engrs Ir.*, 1888, **18**, 66–140.

12. Westland Row Train Shed

HEW 3222
O 166 340

In the days when Ireland had a number of independent railway companies, there was fierce competition to make rail travel as comfortable and convenient as possible for the travelling public. Rail termini needed to be able to handle several trains and many hundreds of passengers at once, as well as freight. Whereas the main station buildings housed such facilities as ticket offices and refreshment establishments, the so-called train shed accommodated the trains drawn up alongside several platforms. In order to afford protection from the elements for the process of boarding and alighting from the carriages, it was necessary to cover these large areas. This presented quite a challenge to the structural engineer and resulted in some fine structures, such as the large curved roof at Westland Row (now Pearse) station in Dublin.

Westland Row was originally the terminus of the Dublin and Kingstown Railway and the earlier train shed was much smaller. The present train shed was completed in 1884 for the Dublin, Wicklow and Wexford Railway.

The roof design is based on that first devised by the famous Dublin iron founder, Richard Turner, for Lime Street station at Liverpool, the plans being obtained from his son William. The principals and arched girders were made in Chepstow, but the rest of the ironwork was supplied by the Dublin firm of Courtney, Stephens and Bailey. The general contractor was Michael Meade & Sons.

The roof of the main bay of the train shed spans 88 ft 10 in., the smaller bay to the south-west having a span of 64 ft 8 in. The top chords of each principal truss are on radii of 56 ft 5 in. and 41 ft 1½ in. respectively. The principals are at 13 ft 4 in. centres and the columns at 40 ft centres. The main roof is 510 ft long, the smaller roof about 240 ft. The principals rest on arched girders spanning longitudinally between the columns.

In 1890–91 the north-west end of the station was further modified to permit trains to connect with Amiens Street (now Connolly) station on the north bank of the river via the Liffey Viaduct. A bowstring girder with curved bottom and top chords was inserted to span the main through platforms and to maintain the structural integrity of the roof. A new street-facing facade was built, consisting of wrought-iron pillars inserted between the existing masonry abutments and a facia formed of decorative cast-iron panels to match the ornamentation on a new bridge spanning Westland Row.

GRIERSON T. B. The enlargement of Westland Row terminus: Part II. *Trans. Instn Civ. Engrs Ir.*, 1893, **22**, 125–147.

13. Heuston Station and Train Shed

The Great Southern and Western Railway (GS&WR), with a total track length of 1150 miles at its peak, operated trains to most of the south and south-west of the country from its Dublin headquarters at Kingbridge (now Heuston) station. The first section of the line as far as Cherryville Junction was opened to traffic on 4 August 1846, shortly after the locomotive works at Inchicore had commenced operations.

HEW 3091
O 138 343

The station consists of an elaborate main building, designed by the English architect Sancton Wood, a pupil of Sir Robert Smirke. It was completed in 1844, and a large train shed to the rear, designed by Sir John Macneill, the Engineer to the GS&WR, was completed four years later.

The main building consists of a central section, two

IARNRÓD ÉIREANN

Heuston station
train shed

storeys high, the lower rusticated, from which spring Corinthian pillars supporting the cornice surmounting the upper storey. Clock towers rise from each of the single-storey flanking wings.

The train shed covers four platforms (Nos. 2–5) and is 616 ft long and 162 ft wide. Seventy-two cast-iron columns with fluted heads, 18 to a row and spaced 35 ft apart, support the longitudinal cast-iron pierced spandrel arches, which in turn support the iron roof trusses of 32 ft span. One platform (the so-called 'military platform' from the days of troop trains) is uncovered. The ironwork was supplied by J. and R. Mallet, the Dublin iron founders. The general contractor was William Dargan, with Cockburn and Williams being responsible for the main building. The two-storey riverside wing was added in 1911. The roof covers an area of about 2½ acres.

Gantries (since removed) were originally used to transfer carriages from one track to another and there were locomotive turntables at the end of each track. The turntables and part of each platform were removed to make way for the present 162 ft by 70 ft passenger concourse.

In November 1996, a major refurbishing project was announced. The main building is to become the passenger entrance with interchange facilities with the projected Light Rail System (LRT) and the existing bus network. Extra platforms are to be provided, but the historic roof structure is to be preserved.

JACOB W. J. Kingsbridge terminus. *Dublin Historical Record*, 1943, **6**, No. 1, 107–117.

14. Inchicore Railway Works

The original buildings at the railway works at Inchicore in Dublin were built between 1844 and 1846 for the Great Southern and Western Railway (GS&WR). The works were in operation from the early part of 1846. Apart from a small works at Limerick (where only freight wagons are now repaired), the Inchicore Works represent the sole survivor of a number of independent railway works in Ireland, and one of the earliest such works surviving in the world.

HEW 3203
O 112 335

The works were built on an 87 acre estate beside the Dublin to Cork main line, about 2 miles to the west of the terminus of the GS&WR at Kingsbridge (now Heuston).

The main building was designed by the architect Sancton Wood, who was also responsible for the headquarters of the company at Kingsbridge. The general contractor was Copthorne.

R. C. COX

Inchicore Railway Works

The original works included a running shed, two erecting shops, boiler, carriage, paint and wagon shops, smithy and foundry, as well as administration and design offices. The roofs of a number of the buildings are supported by iron roof trusses carried on cast-iron columns. These date from the 1840s and were supplied by the Dublin foundry of J. and R. Mallet and are similar to those to be found at Kingsbridge. The foundry has an interesting timber trussed roof from the same date.

The original buildings were built in a very durable blue limestone, the facade towards the railway main line having very distinctive castellated features.

GRAINGER R. The Inchicore works. *Trans. Instn Engrs Ir.*, 1994, **119**.

15. Fairview Rail Bridge

HEW 3095
O 182 363

The problem of constructing arch masonry bridges with centre lines at an angle or skew to the face of the abutments had been tackled by the Romans, but the canal engineer William Chapman is acknowledged to have been the first engineer in modern times to build skew masonry arch bridges in Britain or Ireland. By the time the Dublin to Drogheda railway route was planned in the 1840s, Sir John Macneill was able to specify a substantial skewed masonry rail bridge for the crossing of the Dublin to Howth road east of Fairview.

Although both arches of the bridge appear to have been erected at the same time, the northernmost arch was completed by the contractor Jeffs in 1844, whilst a second arch was added in the 1930s to accommodate road widening, the extra land involved being reclaimed from the sea. The tramway to Howth used to run under the older of the two arches.

The spans are 40 ft on the square and 49 ft 6 in. on the skew, the arches having a rise of 12 ft 6 in. The voussoirs are 2 ft thick and 3 ft deep at the springings, diminishing to 2 ft 6 in. at the crown. The northern abutment and central pier are each 5 ft thick, the southern abutment being 10 ft thick. Six 12 in. thick internal spandrel walls are covered by 6 in. thick granite covers, on which is laid the ballast carrying double tracks, the overall width of the bridge being 30 ft. The approach embankments are contained between 4 ft thick rubble walls.

PATTERSON E. M. *The Great Northern Railway of Ireland.* Oakwood Press, Lingfield, 1962, 21.

16. Liffey Viaduct ('Loop' Line)

The Liffey Viaduct forms part of the one mile 'Loop Line' built under an Act of 1884 for the City of Dublin Junction Railway. This connects the Dublin, Wicklow and Wexford Railway terminus at Westland Row with the Great Northern Railway (Ireland) (GNR(I)) terminus at Amiens Street (now Connolly station), where a new set of platforms was built alongside those of the GNR(I).

HEW 3036
O 167 340 to
O 167 350

The viaduct crossing the River Liffey consists of a three-span bridge of twin wrought-iron latticed girders supported on pairs of cylindrical section cast-iron caissons sunk to rock and infilled with concrete. The spans from the George's Quay side of the river are 116 ft 5 in., 131 ft 2 in., and 139 ft 5 in. respectively.

The simply supported main girders, spaced 30 ft apart, are 13 ft deep with 3 ft wide flanges, the webs being doubly redundant lattices composed of members of varying section. The cross girders originally carried a steel

R. C. COX

Liffey Viaduct

trough deck, but this was replaced in 1958–60 by a system of steel stringers welded to flat deck plating.

The twin river piers are cross braced above high water level by semicircular arched members with hollow spandrels and are capped by ornamental sections, which carry the main girders.

The designer was John Chaloner Smith, and the general contractor M. Meade & Sons of Dublin, the structural ironwork being supplied and erected by William Arrol of Glasgow.

The northern approach viaduct consists of similar type, but smaller, girders carried on white limestone piers (supposedly in an attempt to harmonize with Gandon's Custom House). There are also a number of substantial iron bridges, erected by A. Handyside & Co. of Leeds, spanning the main streets crossed by the line, which was opened on 1 May 1891.

17. Butt Bridge

HEW 3017
O 162 345

Immediately upstream of the Liffey Viaduct, a bridge spans the River Liffey at the point where the Dublin Port and Docks Board erected the Beresford opening swing bridge in 1879. Designed by Bindon Blood Stoney, this bridge had two steep approach spans of 37 ft each and a central swing span of 127 ft, providing two 40 ft navigation openings to allow shipping access to the quays below O'Connell Bridge. The opening span consisted of two parabolic wrought-iron web girders connected by 60 cross girders under the roadway, the girders being carried on a platform which rotated about a central pier.

Following the completion in 1888 of the railway viaduct, it was no longer possible to open the swing span and, with the increase in the volume of road traffic, the narrow and steep roadway soon became a problem.

The present reinforced concrete bridge, named after the Irish parliamentarian Isaac Butt, was designed by Joseph Mallagh (Chief Engineer to the Dublin Port and Docks Board) and Pierce Purcell (Professor of Civil Engineering at University College Dublin) and constructed by Gray's Ferro-Concrete (Ireland) Ltd. Opened in 1932, the bridge is 66 ft wide and carries a 40 ft roadway and two footpaths of 11 ft 9 in. each. The central span of 112 ft is formed by two cantilevered sections, and the two approach spans – each of 39 ft 6 in. – act as counterweights. This is believed to be the earliest recorded use of the cantilevered method of construction in reinforced concrete, albeit on a small scale, in either Britain or Ireland. Over the approach spans the roadway is splayed out to a width of 66 ft at the abutments to give easy access from the quays, the footpaths being carried on ribs cantilevered out from the main structure. The arch profiles are all portions of circular arcs: the central span has a radius of 125 ft with a rise of 13 ft 3 in., while the approach spans have a rise of 8 ft 6 in. and a radius of 27 ft 2 in. on the axis of the bridge, which is slightly skew to the abutments. The parapets are of Wicklow granite.

The removal of around 200 steel sheet piles, employed in the cofferdams surrounding the foundations for the two river piers, was achieved by cutting underwater using oxygen–hydrogen cutting equipment, used here for the first time in Ireland.

BOND F. W. Reconstruction of Butt Bridge, Dublin. *Trans. Instn Civ. Engrs Ir.*, 1933, **59**, 35–116.

R.C COX

Lucan Bridge

18. Lucan Bridge

HEW 3045
O 035 355

The largest single-span masonry arch bridge in Ireland is to be found where the road from Clondalkin to Clonsilla crosses over the River Liffey at Lucan in County Dublin. Many earlier bridges at this spot had been carried away by floods, mostly as a result of undermining of piers and/or abutments.

Lucan Bridge consists of a single arch of ashlar masonry of 110 ft span and 22 ft rise, the arch profile being segmental, with a rise to span ratio of 0.2 (compared with 0.29 at Lismore and similar to that of Sarah Bridge downstream at Islandbridge). The architect and builder George Knowles was probably inspired by the work of Alexander Stevens at Islandbridge. Assisted by James Savage, Knowles created an aesthetically pleasing bridge with raised voussoirs in a deep arch ring and iron balustraded parapets. The parapets were supplied by the Royal Phoenix Ironworks in Parkgate Street in Dublin, their bases being date-stamped '1814'. There is a long approach to the bridge on the south side of the river, the total length of the crossing being around 500 ft. Although there is a weight restriction, the bridge continues to accommodate heavy vehicles from time to time, a good indication of the inherent strength of well-constructed masonry arches.

O'KEEFFE P.J. and SIMINGTON T.A. *Irish stone bridges: history and heritage.* Irish Academic Press, Dublin, 1991, 270–274.

19. Liam Mellowes Bridge

This masonry bridge is the oldest of all the city bridges over the River Liffey still extant, having being built between 1764 and 1768 by Dublin Corporation to replace an earlier bridge swept away by floods in 1763. Designed by the military engineer, General Charles Vallencey, it spans the river between Queen Street on the north bank and Bridgefoot Street on the south. The bridge has three semicircular arches, the central arch having a span of 44 ft 3 in., whilst the flanking arches are each of about 37 ft span. The bridge has some architectural merit with alternate voussoir stones projecting from the arch ring. There are pilasters over the piers and a balustraded parapet. The bridge profile is curved as a consequence of the semicircular arch design and the level of the infill.

HEW 3044
O 145 343

20. O'Connell Bridge

In 1782 the Wide Streets Commissioners obtained a parliamentary grant to erect a bridge to connect Sackville (now O'Connell) Street on the north side of the River Liffey with Westmoreland and D'Olier Streets on the south. The Commissioners employed the architect of the Custom House, James Gandon, to design the bridge. Consisting of three semicircular masonry arches spanning a total of 150 ft between the quayside abutments, this bridge had steep approaches and was quite narrow, being only 40 ft wide. It was named Carlisle Bridge in honour of the Lord Lieutenant of the time.

By the 1860s, the city had spread eastwards along the quays and some 10 000 vehicles a day were using the bridge. Rapidly developing commerce, a railway system largely centred on Dublin and the general economic improvement in the country all combined to make the bridge totally inadequate. Following a design competi-

HEW 3040
O 160 343

tion and many arguments, the proposal of Bindon Blood Stoney, engineer to Dublin port, was finally adopted.

The plan was to rebuild the eighteenth-century structure in order to provide a level roadway and to increase the width to that of Sackville Street. The method of construction involved using concrete filled iron caissons down to rock level to extend the piers to support the new work, together with the rebuilding of the existing arches.

The contractor appointed, William J. Doherty, had already rebuilt Grattan Bridge further upstream and considerable lengths of quay wall. He had also been awarded the contract for the Beresford Swing Bridge. Doherty employed a novel method of forming cofferdams around the piers, driving timber piles in groups in single rows rather than in the more traditional double row pattern. The ironwork for the caissons was supplied by the Skerne Ironworks in Darlington.

The resulting bridge has three elliptical masonry arches: the central arch is of 49 ft span with a rise of 13 ft, and the side arches have a 40 ft span and 12 ft 3 in. rise. The length of the bridge between abutments is 144 ft and the width between parapets 152 ft 8 in. The architectural merits of the bridge lie in the fine granite ashlar masonry work, the balustraded parapets and the sculptured decorations. These carvings by Charles W. Harrison are a reproduction of the original work by Edward Smyth, in particular the carved heads of Neptune and Anna Livia. The bridge was opened in May 1880, when it was renamed in honour of the Irish patriot Daniel O'Connell, whose statue had recently been unveiled nearby.

Cox R. C. *Bindon Blood Stoney: biography of a port engineer.* Institution of Engineers of Ireland, Dublin, 1990, 31–35.

21. Grattan Bridge

HEW 3041
O 154 342

An Act of Parliament in 1811 made the Ballast Board in Dublin responsible for the bridges over the River Liffey upstream as far as and including Victoria (now Rory O'More) Bridge, but excluding the Corporation's Liffey Bridge. This responsibility passed to the Dublin Port and Docks Board when it was established in 1867. Its Chief Engineer, Bindon Blood Stoney, was called upon to un-

R. C. COX

Grattan Bridge

dertake the rebuilding and widening of the Essex and Carlisle (now O'Connell) river bridges.

Essex Bridge was named after the Earl of Essex, Arthur Capel, the Lord Lieutenant of Ireland between 1672 and 1677. This seven-span masonry bridge had been completed in 1678 by Sir Humphrey Jervis and was replaced between 1753 and 1755 by a five arch masonry bridge designed by George Semple. He based his design on Labelye's bridge at Westminster in London. This was the first recorded use in Ireland of cofferdams when laying foundations below water level.

In 1865, traffic conditions required that this narrow bridge with its steep approaches be rebuilt or replaced. Stoney at first suggested an iron bridge, but also put forward plans for rebuilding, which involved substituting a series of flatter segmental or elliptical arches for the lofty crowns of the existing semicircular arches, and carrying the footpaths beyond the face of the arches on cantilevered iron girders.

This latter plan was adopted and the present bridge was completed in 1875 by the contractor W. J. Doherty, with the ironwork supplied by Courtney, Stephens and Bailey of Dublin. The overall span was reduced by widening the quays to accommodate a section of the city's main drainage system, but these low level sewers were later abandoned. The central segmental arch spans 45 ft and is flanked on each side by segmental arches of 39 ft 6 in. span and semicircular arches of 14 ft 2 in. span. The rise to span ratio of the segmental arches is around 0.22. The footpaths are supported on wrought-iron frames cantilevered out from the faces of the bridge and bolted through to the lowest courses of the new masonry. The parapets consist of wrought-iron latticed girders with cast-iron ornamentation. The lamp supports are in the form of sea horses.

Following the rebuilding and widening, the bridge was renamed Grattan Bridge after Henry Grattan (1746–1820). The architectural historian Maurice Craig has described it as 'the ugliest bridge in Dublin', but in engineering terms, Stoney provided an economical solution to the problem set before him.

Cox R. C. *Bindon Blood Stoney: biography of a port engineer*. Institution of Engineers of Ireland, Dublin, 1990, 28–31.

22. Liffey Bridge

HEW 3032
O 158 343

The earliest known iron bridge in Ireland, the Liffey Bridge was erected in 1816 to connect Merchants Arch on the south quays with Liffey Street Lower leading from the north quays. Known variously as the Metal Bridge and the Halfpenny Bridge (after the toll that used to be levied on citizens crossing the River Liffey at that point), the Liffey Bridge is a single span cast-iron arch with an elliptical profile. The bridge consists of three parallel arched ribs spanning 137 ft 9 in. between angled masonry abutments and having a rise of 11 ft 9 in. (The span increases to about 141 ft at deck level). Each arch rib consists of six lengths of cruciform section. These are connected together at each rib joint to form two tiers of rectangular openings with chamfered surround, the depth of the opening decreasing towards the crown. The ribs are stiffened by the deck and by diagonal and normal bracing to form a truss in the plane of the intrados. The transverse cross members are of hollow circular section with a bolt passing through, and act as spacers to provide lateral stability. Cast corbels on the outside ribs carry a

R. C. COX

flat plate which supports the parapet railings. The bridge decking is timber with asphalt surface supported by fish-bellied beams which span perpendicularly to the arch ribs. Three lamps are carried on ornamental supports mounted on top of the railings, their supports providing lateral stability for the railings.

Liffey Bridge

The Liffey Bridge was sponsored by William Walsh and the Lord Mayor of Dublin, John Claudius Beresford. It is said to have been designed by John Windsor, a foreman at the Abraham Darby III foundry at Coalbrookdale in Shropshire. The bridge was transported in eighteen sections down the River Severn to Bristol and thence by sailing ship to Dublin for erection at the site.

DE COURCY J. W. The Ha'penny Bridge in Dublin. *The Structural Engineer*, 1991, **69**, no. 3, 44–47.

23. Rory O'More Bridge

The present bridge spanning the River Liffey between Ellis Street on the north side and Watling Street on the south is the third bridge to be erected on or near the site. The first bridge, erected around 1670, was of timber and replaced a ferry. This bridge was replaced in 1704 with a multi-span masonry bridge, which in turn was replaced

HEW 3039
O 143 343

R. C. COX

Rory O'More
Bridge

by a single-span iron bridge opened in 1861. This is now
known as Rory O'More Bridge (after one of the leaders of
the 1641 rebellion), but was originally named Victoria
Bridge.

The bridge has seven segmental cast-iron ribs of I-section at 5 ft 6 in. centres, spanning 95 ft on a slight skew
between granite ashlar abutments at the quay walls, the
rise of the arches being 9 ft 6 in. The ribs are 24 in. deep
with 9 in. flanges and there is wrought-iron diagonal
bracing between the ribs. The carriageway is 23 ft wide
with 5 ft wide footpaths each side. The decking is carried
on buckled plates spanning between I-beams placed on
top of the spandrels, which are composed of a number of
vertical members connected at their tops by semicircular
arched elements. The arches spring from masonry abutments.

The ironwork was supplied by Robert Daglish Junior
of St Helen's Foundry in Lancashire, the general contractor being John Killeen of Malahide. According to the
inscription on the arch, the ironwork was cast in 1858,
although the bridge was not opened until some three
years later. The bridge was commissioned by the Corporation for Preserving and Improving the Port of Dublin

and the overall design and construction was the responsibility of the port engineer, George Halpin (Senior).

24. Sean Heuston Bridge

Spanning the River Liffey near Heuston (formerly King's Bridge) station is an iron road bridge erected in 1828 to commemorate the visit of George IV to Ireland seven years previously. The bridge, formerly called King's Bridge, but now named after an Irish patriot, Sean Heuston, consists of a single cast-iron span of 98 ft with a rise of 16 ft 10 in. There are seven ribs, the outer ones, together with the spandrels and parapets, being highly ornamented.

HEW 3033
O 138 343

The bridge was designed by the architect George Papworth and the castings produced by Richard Robinson at the Royal Phoenix Iron Works in nearly Parkgate Street. The bridge has an overall width of 31 ft 2 in. and, since 1980, has carried a 2 ton weight restriction. Following the opening of the nearby Frank Sherwin Bridge, the iron bridge serves as a one-way access for light vehicular traffic to Heuston station. The bridge is to be used to carry Dublin's projected LRT system across the river here.

Sean Heuston Bridge

R. C. COX

25. Glasnevin Curvilinear Glasshouses

HEW 3201
O 152 372

Situated on the banks of the River Tolka at Glasnevin in the north of Dublin city, the curvilinear range of glasshouses at the National Botanic Gardens is probably the finest example of Victorian building extant in Ireland and one of the most important nineteenth-century glasshouses surviving in Europe. The entire range has recently (1992–95) been restored by John Paul Construction, working under the direction of the Project Architect, Ciaran O'Connor of the Office of Public Works, and drawing on the structural engineering experience of Ove Arup & Partners. The contractors, employing many skilled artisans, painstakingly dismantled the entire structure and, using only original or contemporary materials, carried out a thorough restoration. Through the development of new conservation techniques, 87 per cent of the original wrought-iron glazing bars have been restored, the remainder being made up of reforged old wrought iron, including pieces from the Kew Gardens Palm House (replaced in stainless steel in the 1980s). Similarly, all cast-iron work, such as the decorative columns, were faithfully restored.

The glasshouses were erected on a phased basis between 1843 and 1869, the east wing being the first to be completed by William Clancy to the design of Duncan Ferguson. This was followed in 1845–46 by the central house and the west wing and connecting corridors, constructed by the famous ironmaster Richard Turner to the detailed designs of Frederick Darley, although the conceptual design is generally attributed to Turner. In 1869, Richard Turner's son, William, planned and constructed major extensions to the east and west wings and added turrets.

The overall length of the curvilinear range is 333 ft and the height of the central house 33 ft 9 in. The wings are 35 ft 8 in. wide and 17 ft 4 in. high.

In 1997, the Office of Public Works was awarded the prestigious Europa Nostra medal for the 'excellent and faithful restoration of one of the most important surviving nineteenth-century glasshouses in Europe'.

O'CONNOR C. *Restoration of curvilinear range—National Botanic Gardens.* Office of Public Works, Dublin, 1995.

R. C. COX

26. Liffey Tunnel

Glasnevin
Glasshouses

HEW 3251
O 181 342

The first of many subaqueous tunnels to be built was the tunnel under the River Thames in London, constructed between 1825 and 1843 by the Brunels (father and son). Ireland's first and only subaqueous bored tunnel was not constructed until a century later, driven under the River Liffey in Dublin in the 1920s.

The tunnel is 831 ft in length between access shaft centres, 7 ft in diameter, and conveys water and drainage pipes. It was designed by the City Engineer, Michael A. Moynihan and constructed by Sir Robert McAlpine & Sons (Ireland).

The 107 ft deep south shaft is located near the river wall at the end of Thorncastle Street in Ringsend and the 112 ft deep north shaft at the commencement of the North Wall quay extension (across the East Wall road from the Point Depot). The shafts are 11 ft in internal diameter and were constructed of 11 ft 8 in. deep cast-iron rings lined with concrete.

The tunnel was driven through rock, mostly limestone, and lies about 60 ft below the deepest part of the river bed. Considerable problems were encountered during construction because of water entering the tunnel and shafts and this led eventually to compressed air working. By 1928, the project had taken two years to complete. Twenty years earlier, some 2½ miles of 8 ft diameter tunnel was driven from Pigeon House sewage works to Burgh Quay as part of Dublin's Main Drainage Scheme.

NICHOLLS H. The construction of a tunnel under the River Liffey. *Trans. Instn Civ. Engrs Ir.,* 1928–29, **55**, 186–230.

27. Donnybrook Bus Garage

HEW 3224
O 178 313

The main building at the bus garage at Donnybrook in Dublin was completed in 1951 and is covered by the most extensive area of concrete shell roofing in Ireland. Columns support a roof of ten thin slab vaults. The width of the structure is 110 ft and the length about 400 ft, divided into two five-bay sections. Each vault is 40 ft wide and spans 104 ft between the main columns. The height to the underside of the longitudinal beams is 20 ft. The vaulted roof springs from a point 26 ft 2 in. above floor level and the highest point of the roof is 33 ft 6 in. Roof lights 7 ft 6 in. wide extend the full length of the vaults. The external

Donnybrook Bus
Garage

R. C. COX

walls are secured to the columns by flexible ties to give the roof complete freedom to expand longitudinally.

The vaulted roof slabs are only 2¾ in. thick, splaying out to 5 in. for a short distance at the springing, at the top light trimmer beams, and at the junction with the gable walls. There are five layers of reinforcing steel of varying diameter with ½ in. cover top and bottom.

The architects were Michael Scott and James Brennan and the structural design was by Ove Arup & Partners. The contractors were H. C. McNally & Co., in collaboration with Messrs Larsen and Neilsen A.S. of Copenhagen.

ANON. Omnibus station and garage, Dublin. *Conc. and Constr. Eng.*, 1950, **45**, no.3, 87–96.

28. Balrothery Weir

As early as the thirteenth century, efforts were being made to ensure a better and more reliable water supply to Dublin city. At that time, the main source of supply to those citizens living within the city walls was from the River Poddle (now entirely culverted). The somewhat limited catchment area of the Poddle, a tributary of the River Liffey, lies to the west of Tallaght. In the thirteenth

HEW 3229
O 113 277

Balrothery Weir

R.C.COX

century, the monks of the Priory of St Thomas constructed a weir across the River Dodder at Balrothery near Firhouse, at which point they abstracted a water supply for the priory.

In order to increase the flow in the River Poddle, and so improve the water supply to Dublin, the weir at Balrothery was raised in 1244 and water diverted through an open artificial channel some 1¾ miles long to the Tymon river upstream of Kimmage Manor at Whitehall Road (from here the Tymon river formed a natural channel to the River Poddle). The flow in the canal was controlled by four sluice gates. The weir, built of rough limestone blocks, was reconstructed by Andrew Coffey in the early nineteenth century, the present sluice gates and bypass channels dating from that time. Major repairs were also required following damage caused by Hurricane Charlie in 1986. The weir is about 230 ft wide and 16 ft in height.

SWEENEY C. L. *The rivers of Dublin*. Dublin Corporation, Dublin, 1991, 29.

29. Vartry Aqueduct and Stillorgan Reservoirs

**HEW 3022
(Aqueduct) and
HEW 3023
(Reservoirs)**

**O 215 014 to
O 200 268**

The purpose of service reservoirs is to provide water storage as near as possible to the point of usage, in this case the southern suburbs of the city of Dublin and the towns lying along the coast to the south. The original Vartry Water Supply Scheme, built between 1862 and 1867, included one large service reservoir at Stillorgan at the end of an aqueduct from Roundwood in County Wicklow. The aqueduct is 33 miles long. It begins at O 215 014 as a 6 ft high by 5 ft wide tunnel, which was driven in solid rock for 2½ miles, and continues as a twin pipeline.

A tunnelling machine was used for driving part of the tunnel and was subsequently used for exploratory work for the Channel Tunnel in England by its inventor, Colonel Beaumont.

The aqueduct is carried over the Dargle (O 238 165) and Cookstown (O 235 171) rivers on ornamental iron latticed truss bridges of 65 ft and 45 ft span respectively. Stillorgan is about 4½ miles from the Grand Canal at

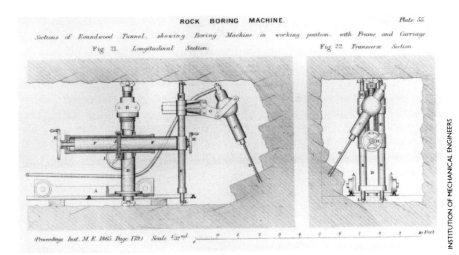

ROCK BORING MACHINE. *Plate 55.*

Sections of Roundwood Tunnel, showing Boring Machine in working position, with Frame and Carriage.

Fig. 21. *Longitudinal Section.*   Fig. 22. *Transverse Section.*

(Proceedings Inst. M.E. 1865. Page 179.) Scale 1/32nd   10 Feet

Leeson Street Bridge, where there is a siphon over the canal.

A valve house and screen chamber were built at the north-east corner of the reservoir, but these are now disused. The extension of the Vartry supply during the period 1908 to 1925 included the provision of two new reservoirs at Stillorgan and a new valve house.

The Stillorgan reservoirs were created partly by excavation, partly by earthen embankments, the highest embankment being about 26 ft above the general ground level. The Lady Gray Reservoir at Stillorgan, completed in 1923, was named after the wife of Sir John Gray, the Chairman of the 1860s Waterworks Committee, and has a capacity of 90 million gallons.

The present valve house, which contains the screen chamber, is an octagonal shaped building with sides of 18 ft. The building rises about 15 ft above ground and extends to 50 ft below ground. There are some fine decorated iron trusses supporting the roof.

The design of the original scheme was the responsibility of the Dublin City Engineer, Parke Neville; the pipework, valves and other ironwork were supplied and erected by Edington & Son of Glasgow.

NEVILLE P. On the water supply of the city of Dublin. *Min. Proc. Instn Civ. Engrs*, 1874, **38**, 1–49.

LOW G. Description of a rock boring machine. *Proc. Instn Mech. Engrs*, 1865, 179–200.

Rock boring machine

O 200 268

30. UCD Water Tower

HEW 3240
O 178 312

A very striking reinforced concrete water tower was built on the Belfield campus of University College Dublin (UCD) in 1969–70. The tower was provided to ensure an adequate supply at a pressure higher than that which could, at the time, be provided from the public mains. The tower is no longer used for this purpose.

The top of the tower is 184 ft above ground level. The shaft is a regular pentagon with an outer side length of

UCD Water
Tower

R. C. COX

7 ft 6 in. The wall thickness is 20 in. for just over the first third, when it reduces to 14 in. The 150 000 gallon tank is an asymmetrical pentoid with a maximum width of 41 ft 6 in. The original design comprised a dodecahedron with twelve pentagonal faces, but the lower part had to be modified to allow for the pipework and for merging into the base. The tank walls are also 14 in. thick.

In line with one side of the main shaft, a 5 ft by 3 ft access shaft passes through the tank to the flat roof. The whole shape has been emphasized by 2 in. deep fluting. The base comprises a 5 ft thick 44 ft square reinforced concrete raft.

The architect was Andrew Wejchert, the structural design was by Thomas Garland & Partners, and the contractor was Paul Construction Ltd of Dublin.

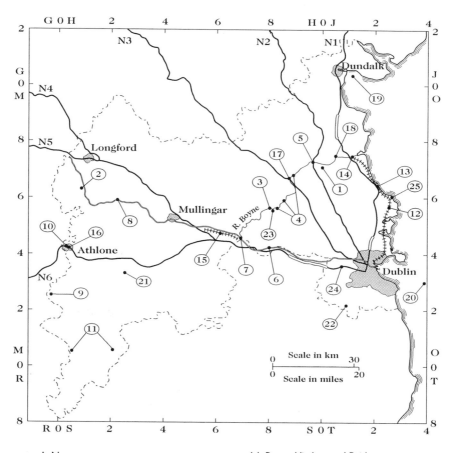

1. Newgrange
2. Corlea Trackway
3. Trim Bridge
4. Early Boyne Bridges
5. Slane Bridge
6. Royal Canal
7. Boyne Aqueduct
8. Whitworth Aqueduct
9. Shannonbridge
10. Athlone Road Bridge, Weir and Locks
11. Birr and Kinnity Suspension Bridges
12. Dublin and Drogheda Railway
13. Balbriggan Viaduct

14. Boyne Viaduct and Bridge
15. Railways across Bogs
16. Athlone Rail Bridge
17. Navan Viaduct
18. Obelisk Bridge
19. Dundalk Bay Pile Light
20. Kish Bank Lighthouse
21. Goodbody's Chimney, Clara
22. Bohernabreena Reservoirs
23. Trim Water Tower
24. West Link Bridge
25. Skerries Windmill

2. North Leinster (except Dublin City and District)

Leinster is the largest of the ancient provinces of Ireland. For the purposes of this book, the province has been divided into two regions. North Leinster comprises the counties of Meath, Louth, Longford, Westmeath, Offaly and Dublin (now divided into the administrative areas of Fingal, South Dublin and Dun Laoghaire-Rathdown). The remaining counties in the province are grouped together as South Leinster.

North Leinster extends from the east coast as far west as the River Shannon, northwards to the Carlingford peninsula and south to the southern fringes of the central plain. The area is for the most part flat and low-lying, and is characterized by a mixture of peat bogs, lakes, fertile farmland and slow moving rivers.

Outside the Dublin region, agriculture has always been the main activity, with some other industries being established in more recent times, notably the mechanized extraction of peat for use as a fuel in power stations.

The River Shannon forms a natural barrier to communication between Leinster and Connaught and its associated civil engineering heritage is dealt with in Chapter 6. The River Boyne, and its main tributary, the Blackwater, flow through what was once the ancient province of Royal Meath. Some of Ireland's oldest masonry arch bridges were erected across these rivers to accommodate ancient routes and much later the turnpike roads.

Road, canal and railway engineers had to deal with the problem of the instability of large areas of bog, or were forced to plan lengthy detours. The Midland Great Western Railway engineers carried out extensive drainage works before laying down their permanent way. The rail bridge across the Shannon at Athlone, erected in 1851, is also a monument to these pioneering railway engineers.

The Royal Canal traverses North Leinster and is currently undergoing a complete restoration of the navigation. There are fine aqueducts at

Longwood in County Meath and at Abbeyshrule in County Longford. An early reinforced concrete water tower near Trim, dating from 1908, is still in service.

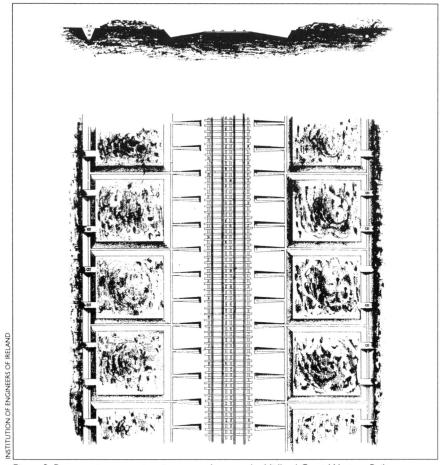

Figure 3. Permanent way construction across bogs on the Midland Great Western Railway

1. Newgrange

Newgrange in County Meath is one of the finest examples in Western Europe of a type of tomb known as a passage grave. The tomb consists of a passage and central chamber, the walls and roof of which are constructed of large slabs without mortar. A large circular mound or cairn of stones covers the tomb and a kerb of massive slabs laid on their long edges, ends touching, surrounds the base of the cairn and acts as an earth retaining wall.

HEW 3217
O 006 728

The diameter of the mound varies from 259 ft to 278 ft. The chamber interior is 17 ft by 21 ft 4 in. by 19 ft 6 in. in height. This is reached by a passage, 62 ft long and averaging around 5 ft high.

Newgrange

DEPT OF ARTS, CULTURE AND THE GAELTACHT, IRELAND

The Stone Age people who settled in the Boyne valley around 3000 BC were a farming and stock-raising community, but displayed considerable architectural and engineering skills in the completion of the structure at Newgrange.

O'KELLY C. *Newgrange, County Meath*. Office of Public Works, Dublin, 1982.

O'FLAHERTY E. *Newgrange*. Clashganna Mills, Borris, County Carlow, 1980.

2. Corlea Trackway

HEW 3261
N 086 629

At Corlea bog in County Longford, the industrial exploitation of the peat has uncovered (and mostly destroyed) samples of an Iron Age timber trackway, which has been dated to 148 BC. Overlain are later replacement trackways, the most recent dating from around 600 AD.

Split oak sleepers were laid transversely on long straight runners. In places, pegs held the runners in position on the bog. Excavation has shown older timbers, which appear to have been reused, below the road. Most appear to be parts of flat-bottomed tub-like vessels.

The exact purpose of this massively constructed trackway is unclear, but, like Newgrange, it is evidence that a strong communal grouping was at work at this time.

A 60 ft length of what is said to be the largest trackway of its type in Europe has been preserved on permanent display. It is protected in a building with a controlled atmosphere. Around this, a block of original bog has been wrapped in plastic sheets to prevent the site and the timbers from drying out.

RAFTERY B. Ancient track-ways in Corlea bog. *Longford Archaeological Journal*, 1988, 60–64.

3. Trim Bridge

HEW 3157
N 801 570

This ancient masonry bridge over the River Boyne at Trim in County Meath is situated near the castle and consists of four pointed segmental masonry arches of 16 ft span with a rise of 6 ft 9 in. The arch rings are formed of roughly trimmed, rectangular stones of varying thickness (3 in. to 6 in.). The 8 ft thick piers and abutments are

R. C. COX

founded on rock, and the bridge is 21 ft wide with solid   Trim Bridge
parapets finished with coping stones.

A great flood in 1330 destroyed all the bridges on the
Boyne, with the exception of Babe's, and the present
bridge was most likely built in 1393. It is probably the
oldest unaltered bridge in existence today in Ireland.

During the 1970s, the Office of Public Works carried
out an arterial drainage scheme, which involved lowering the bed of the river at this point by 4 ft. As with many
other bridges affected by the scouring action of rivers, the
base of each pier of the bridge was surrounded with a
reinforced concrete skirt, incorporating pointed cutwaters to assist the flow of flood waters under the bridge.

O'KEEFFE P. J. and SIMINGTON T. A. *Irish stone bridges: history and heritage.*
Irish Academic Press, Dublin, 1991, 151–153.

4. Early Boyne Bridges

The River Boyne winds its way at a leisurely pace through
the rich grasslands of County Meath. A number of ancient north–south routes are carried across the river on
masonry arch bridges of varying antiquity. More is
known about the origins of some than others. All were

R.C.COX

Bective Bridge

HEW 3158
N 815 568

HEW 3159
N 886 656

HEW 3160
N 859 598

underpinned by the Office of Public Works during the course of an arterial drainage scheme in the 1970s.

Newtown Bridge, to the east of the town of Trim, was reputedly sponsored by William Sherwood, Bishop of Meath, and erected between 1460 and 1475. The abutments and piers are founded on rock and this fact has probably contributed to the ability of the structure to withstand above average flood conditions. The minimum width of the bridge is only 13 ft and it appears never to have been widened. The five arch bridge is slightly curved in plan towards the current and the roadway is humpbacked. The upstream cutwaters are massive triangles of solid masonry capped by an inverted triangle, a new type of design for the period.

Kilcarn Bridge, two miles south of Navan, was probably erected in the late sixteenth century and widened shortly after 1729, when the Dublin to Navan road became a turnpike. Around 1800, six of the eleven arches suffered flood damage and were replaced by the present four larger spans. The bridge has been maintained in good repair, and is now bypassed.

Bective Bridge is a multi-span masonry bridge crossing the Boyne near Bective Abbey, about six miles to the south of Navan. The profiles of the eleven arches are virtually semicircular, the spans varying from 10 ft to 14 ft, with pier thicknesses of 4 ft to a maximum of 7 ft 6 in. The third cutwater from each bank is brought up to form pedestrian refuges. Bective Bridge was probably erected in the second half of the seventeenth century, making it the youngest of the three described here.

O'KEEFFE P. J. and SIMINGTON T. A. *Irish stone bridges: history and heritage*. Irish Academic Press, Dublin, 1991, 165–166, 180–183, 217–218.

5. Slane Bridge

The multi-span masonry road bridge spanning the River Boyne near Slane Castle in County Meath is built in a mixture of styles. The pointed segmental arches to be seen on the downstream side indicate that the original parts of the structure could date from the mid-fourteenth century. According to O'Keeffe, the bridge, erected for the Fleming family, was widened some time before 1600.

HEW 3151
N 963 736

Of the 13 arches in the overall bridge length of 500 ft, the most southerly was added around 1750 in order to accommodate the passage of boats on the Boyne Navigation. There are then two flood relief arches before the eight arches spanning the main waterway. On the north-east side, two arches were added in 1776, when a weir

Slane Bridge

R. C. COX

and mill-race were built. The main river arches vary in span from 12 ft 6 in. to 21 ft.

The bridge was further widened by 8 ft on the upstream side, probably around 1800, the arch profiles here being semicircular. The present overall width is 24 ft, or 21 ft between parapets. This narrow ancient bridge carries heavy traffic on the main road between Dublin and Monaghan.

O'KEEFFE P. J. and SIMINGTON T. A. *Irish stone bridges: history and heritage.* Irish Academic Press, Dublin, 1991, 153–156.

6. Royal Canal

HEW 3002
O 173 345 to
N 060 760

The Royal Canal Company was incorporated in 1789 to build a canal to link Dublin with the River Shannon at Cloondara in County Longford. Work commenced in 1790 on the 90 mile long main line, but progress was slow and it eventually took 27 years before the link could be completed. The Royal Canal, unlike its rival Grand Canal, did not offer passenger and freight services between Dublin and Limerick and had to be content with serving the smaller centres of population in the North Midlands and the Upper Shannon.

The line had originally been surveyed by John Brownrigg, the engineer to the Directors General of Inland Navigation, but the direction of the actual route and the detailed engineering design became the responsibility of the engineer to the canal company, Richard Evans. However, he died in 1812. As a result of financial difficulties, the construction of the canal was taken over in 1813 by the Directors General of Inland Navigation and was completed to the Shannon by 1817 under the direction of their engineer, John Killaly. The contractors were David Henry, Bernard Mullins and John MacMahon, destined to become one of the leading contracting firms in Ireland in the early nineteenth century. Bernard Mullins left his valuable library to the Institution of Civil Engineers of Ireland and money to enable a medal to be offered in perpetuity for a paper presented at a meeting of the Institution.

The canal commences at a sea lock at the North Wall quays in Dublin and passes through the Spencer Dock, built in 1875 by the Midland Great Western Railway

(MGWR). The canal is carried over the Rye Water near Leixlip in County Kildare on a massive embankment, the river flowing underneath in a 230 ft long tunnel, 30 ft wide with a semicircular arched roof. Major aqueducts were constructed at Longwood in County Meath over the River Boyne and at Abbeyshrule in County Longford over the River Inny, and smaller ones at a number of other locations. The summit level of the canal ends at Coolnahay, some six miles beyond Mullingar, which was reached in 1806. There are a total of 46 locks on the canal, 10 of these being double-chambered.

The canal company was purchased in 1845 by the MGWR, who built their line alongside the canal from a Dublin terminus at Broadstone to a few miles west of Mullingar. Ownership of the canal transferred from Córas Iompair Éireann (the National Transportation Company) to the Office of Public Works in 1986 and a vigorous programme of restoration was begun. The navigation has now been restored between Dublin and Mullingar and restoration as far as the Shannon is in progress.

Dry dock at Richmond Harbour on the Royal Canal

CLARKE P. *The Royal Canal: the complete story.* Elo Publications Ltd, Dublin, 1992.
DELANY R. *Ireland's inland waterways.* Appletree Press, Belfast, 1992, 98–108, 145–149.

7. Boyne Aqueduct

HEW 3128
N 692 453

The Boyne Aqueduct was completed in 1804 and carries the Royal Canal over the River Boyne near Longwood in County Meath. The river at this point is about 40 ft wide and flows through the central arch of the aqueduct.

The designer, Richard Evans, was faced with the problem of providing a structure capable of withstanding not only the forces applied by the canal waters, canal traffic and the dead weight of the masonry, but also the considerable lateral forces resulting from the river when in flood. The resulting viaduct in rusticated ashlar limestone masonry has three arches spanning the river and its flood plain.

The arch profiles are three-centred and are each about 40 ft span with a rise of about 16 ft. The voussoirs in the arch rings are 2 ft deep and the tops of the spandrel walls terminate in a cornice running the length of the aqueduct. The 6 ft thick piers have a simple architectural feature rising from curved cutwaters. The massive abutments are splayed at the base to a width of 107 ft 6 in.

This stretch of the canal is about 35 ft wide and narrows to 20 ft over the aqueduct, the total length of which is 132 ft. There are 8 ft 9 in. wide towpaths on each side of the waterway.

The main Dublin to Galway railway line runs alongside the canal and crosses the river on a parallel masonry arch viaduct, built in 1849, which has three flat elliptical arches of rise to span ratio of around 0.25 and of similar span to those of the aqueduct.

8. Whitworth Aqueduct

HEW 3124
N 232 600

The Whitworth Aqueduct carries the Royal Canal over the River Inny near Abbeyshrule in County Longford. It was designed by the talented Irish canal engineer, John Killaly, and constructed between 1814 and 1817 by the partnership of Henry, Mullins and MacMahon.

The aqueduct consists of five segmental arches, each spanning 20 ft with a rise of 5 ft. The width of the structure at the springing level of the arches is 38 ft 9 in., but this reduces to 35 ft at the level of the canal. The arch extrados

R. C. COX

is formed of cut limestone voussoirs which project from the face of the spandrel walls. The river piers and splayed abutments are carried up from river level in ashlar limestone detailing to a cornice. The spandrel walls are of rough hewn coursed limestone, as are the parapet walls, which extend between decorative panels over the piers.

Whitworth Aqueduct

The overall length of the aqueduct is 165 ft and the overall width 35 ft at the canal level. There are towpaths 8 ft wide on both sides of the waterway, which is 15 ft 6 in. at the water-line reducing to 13 ft at the bottom.

DELANY V. T. H. and DELANY D. R. *The canals of the south of Ireland.* David and Charles, Newton Abbot, 1966, 77 ff.

9. Shannonbridge

As part of the improvement of the Shannon Navigation in the 1840s, all the major bridges crossing the river were either replaced or altered by the inclusion of an opening span to permit navigation by steamers plying between Limerick, Athlone and the terminal harbours of the Grand and Royal Canals from Dublin.

HEW 3132
M 968 255

At Shannonbridge it was decided to retain the eighteenth-century multi-span masonry arch bridge, but to underpin the foundations and provide a twin-leaf cast-iron swivel bridge over the navigation channel at the eastern end of the crossing.

The opening span was designed by Thomas Rhodes, Chief Engineer to the Shannon Commission, and manufactured and erected in 1843 by the Dublin firm of J. and R. Mallet. Individual castings for the swivel bridge, weighing anything from 10 to 18 tons, were transported from Dublin to the Shannon by road and canal. The swivel bridge was replaced in the 1980s by the present fixed beam and slab arrangement spanning the 65 ft opening, the original bridge castings being preserved in an adjacent car park.

The bridge has an overall length of 692 ft and contains a total of 16 spans (excluding the navigation span), 14 of 20 ft span, and two end spans of 15 ft. The arches are all semicircular and span between 6 ft thick piers. The bridge is only 16 ft wide between parapets and traffic is single lane and controlled by lights.

There is fine example of a bridge master's house at the eastern end of the bridge, designed by Thomas Omer around 1760 as part of the earlier improvements to the navigation. There are extensive fortifications at the western or Connaught end of the bridge.

10. Athlone Road Bridge, Weir and Locks

HEW 3180
N 039 416
(Bridge)

Athlone was, from earliest times, the principal crossing point of the River Shannon between Leinster and Connaught, and was heavily defended. It is on the main route from Dublin to Galway.

Records indicate that, between 1120 and 1159, one or more timber bridges had existed on or near the site of the present bridge. The first masonry bridge at Athlone was erected around 1210. This was rebuilt in 1275 and replaced by a ferry in 1306.

Sir Henry Sidney commissioned a nine or ten arch masonry bridge, which was completed in 1567, the engineer being a Peter Lewys. A system of tolls operated up until 1826.

Athlone Bridge at this time was only 14 ft in width and the Shannon Commission's engineer, Thomas Rhodes, recommended its replacement by the present masonry arch bridge, each arch being of 61 ft span and elliptical in form. This was built between 1841 and 1844 by the contractor John MacMahon. This was the first recorded occasion in Ireland on which complete arch centrings were eased down and reused, MacMahon having purchased the centrings used for the bridge further upstream at Banagher, which had 60 ft spans.

A 40 ft cast-iron swivelling span at the western end was replaced in 1962 with a fixed reinforced concrete slab. Through traffic on the national primary route is now carried across the river on the new Shannon Bridge, opened in 1980 as part of the Athlone bypass.

Immediately south of the Athlone Bridge are the large

Athlone Bridges, Lock and Weir, on the Shannon Navigation

weir and locks of the Shannon Navigation. The old by-pass canal to the west was abandoned in the 1840s and a new channel excavated in the river bed. A new lock was built in the dry inside a large caisson. Caisson work had also been necessary for the construction of the bridge piers. Major difficulties occurred when constructing the weir, which became undermined. To allow work to proceed, the river was diverted down the old canal and the weir was rebuilt, not without problems occasioned by failure of the temporary dam.

KEARNEY M. and O'BRIEN G. (eds). *Athlone: bridging the centuries*. Westmeath County Council, Mullingar,1991.

11. Birr and Kinnity Suspension Bridges

HEW 3242

Birr Castle, beside the town of Birr in County Offaly, is the seat of the Earls of Rosse and the home of the Parsons family. The Camcor river, a tributary of the Little Brosna river, flows through the grounds of the castle.

N 053 048

Some time before 1826, William Parsons, the 2nd Earl of Rosse, commissioned the erection of a small iron suspension bridge of 44 ft span across the Camcor. Thomas Cooke, writing in 1826, describes the bridge as ' a curious wire bridge'. Exact dating remains a problem but, following a fire in 1823, the castle was enlarged and this may well have been the time when the bridge was erected, although an earlier date is possible. There are enough surviving elements of forged iron, typical of the blacksmith's art in the period 1810 to 1830, to imply that the cables and the oldest of the suspenders are original. This would make the bridge at Birr, in all probability, the oldest surviving wire suspension bridge in Europe.

The 34 in. wide timber deck is carried on 4 in. by 2 in. timber stringers, which in turn are suspended by vertical wire hangers from a pair of cables. The suspension cables consist of eight ¼ in. diameter forged iron wires, and are fixed at each end of the bridge to portal frames at a height of 5 ft 6 in. above deck level. The portal frames are tied back to ground anchorages in the masonry abutments. The bridge has forged iron latticed side panelling, typical of such bridges of the period. The portal frames were

M. B. BARRY

subsequently encased in concrete, probably around 1911, and this precludes an examination of the underlying original material.

However, a similar bridge of smaller span was located at Kinnity Castle, spanning the Camcor about 12 miles upstream. This has not been modified and is identical in design to the bridge at Birr, with the exception of the span, which is only 26 ft 6 in. The portals are formed from pairs of 4 in. diameter cast-iron tubing, connected across the top and suitably braced. The name of the maker stamped on the Kinnity portals is T. and D. Roberts.

COOKE T. L. *Picture of Parsontown*. W. de Veaux, Dublin, 1826, 170.

Birr Castle
Suspension Bridge

N 203 055

12. Dublin and Drogheda Railway

Following the success of the Dublin and Kingstown Railway, a Bill put forward by the promoters of a line of railway northwards from the capital received royal assent on 13 August 1836. A route between Dublin and Drogheda had been surveyed by the English engineer, William Cubitt, and work commenced in 1838 at Kilbarrack on the section between the Royal Canal and Portmarnock, W. R. Weeks being the contractor.

HEW 3098
O 166 350 to
O 100 748

Sir John Benjamin
Macneill,
Engineer to the
Dublin and
Drogheda Railway

TRINITY COLLEGE DUBLIN

As no gradient was to exceed 1 in 160, in order to accommodate the locomotives of the day, some heavy engineering was required, such as the cuttings at Malahide and Skerries and the embankment at Clontarf near Dublin. However, in the light of a recommendation contained in the Report of the Railway Commissioners (1838) that the line to the north should go through Navan and Armagh, rather than follow a coastal route, work was brought to a halt.

In February 1840, John Macneill, one of the Commissioners, was appointed Engineer to the Dublin and

Drogheda Railway Company and a new Bill was expertly steered through Parliament by Daniel O'Connell. A new contract was awarded to Messrs Jeffs, who completed the Killester cutting and Clontarf embankment. The remainder of the line to Drogheda was contracted to William Dargan and William McCormick.

The line was opened on 24 May 1844 by the Lord Lieutenant and Macneill received a knighthood for his services to the community. Amiens Street (now Connolly) station was completed two years later, but was later extended. The main building, designed by William Deane Butler, is in the Italianate style, then new to Dublin, and was built of Wicklow granite from the Goldenhill Quarries.

The station is at a high level and the tracks are carried northwards on a viaduct of 75 arches as far as the Royal Canal. To cross the canal, Macneill used iron latticed girders of 144 ft span, which were the largest of their type in the world at the time of erection. These lasted until 1862, when increasing loads required an intermediate pier to be erected in the bed of the canal. The latticed girders were replaced by Pratt-type trusses in 1912.

The original timber viaduct at Malahide was replaced in 1860 by an iron structure and this in turn between 1966 and 1968 by the present concrete viaduct. By comparison, the masonry viaduct at Balbriggan is as originally constructed by William Dargan in 1843. The other viaducts over the river estuaries at Rogerstown, Gormanstown and Laytown were also originally in timber, but were replaced in wrought iron in the 1880s, and were further strengthened in 1996 as part of the upgrading of the Dublin–Belfast rail link. A branch to Howth was completed in 1847. The line became part of the Great Northern Railway (Ireland) network in 1876.

GAMBLE N. E. The Dublin and Drogheda Railway, 1844–1847: Parts I, II and III. *Journ. Ir. Rlwy Rec. Soc.*, 1981–1982, **14**, no. 84, 162–170; no. 85, 228–235; no. 87, 318–325.

MURRAY K. A. *The Great Northern Railway (Ireland): past, present and future.* GNR(I), Dublin, 1944, 14–36.

13. Balbriggan Viaduct

HEW 3035
O 202 640

At Balbriggan, at the northern extremity of County Fingal between the town and a small fishing harbour, the Dublin and Drogheda Railway is carried over four roads and a small river on an eleven arch viaduct built in 1843–44 by the contractor William Dargan to the design of Sir John Macneill, Engineer to the railway company.

Each of the segmental arches is of 30 ft span with a rise of 10 ft. Brick arch rings with internal brick spandrel walls span between 6 ft thick brick piers, the outer arch rings and spandrel walls being of limestone. The individual voussoirs are 3 ft thick and 2 ft deep. The total length of the viaduct between the 10 ft thick abutments is 390 ft. The rails are carried on 8 in. thick spandrel covers supported on seven diaphragm walls, which are alternately 15 in. and 12 in. thick.

Footpaths, each 6 ft 6 in. wide, are carried at rail level on each side of the viaduct on cast-iron arches spanning between the 41 ft wide stone-faced brick piers. Cast-iron pilaster caps over each pier are connected by iron railings along the length of the viaduct. Some of the arches were bricked in at a later date in order to block the glow from nearby railway coke ovens. This was done to prevent

Balbriggan Viaduct

R. C. COX

confusion to shipping approaching the Boyne estuary further north.

14. Boyne Viaduct and Bridge

The imposing viaduct and bridge over the estuary of the River Boyne at Drogheda, about 32 miles north of Dublin, was constructed in its original form between 1851 and 1855 as the final section of the Dublin and Belfast Junction Railway, linking the Dublin and Drogheda Railway at Drogheda with the Ulster Railway at Portadown.

HEW 3013
O 098 754

The Admiralty, as at the Menai Straits in Wales, insisted on minimum headroom of 90 ft over a minimum clear waterway of 250 ft. To meet these conditions, Sir John Benjamin Macneill, having introduced the iron latticed girder bridge to these islands in 1843, suggested that such a bridge would be suitable for the main span of the Boyne crossing, but left the details of the design to James Barton, Chief Engineer to the railway company. The latticed girders were designed to be continuous across the supports. The principles of multi-lattice construction in wrought iron were first applied on a large scale in the Boyne Viaduct. This led to a significant increase in knowledge of the structural properties of wrought iron and encouraged more exact computation of the stresses in structural components.

Barton was one of the first two civil engineers to graduate from the Trinity College Dublin School of Civil Engineering (founded in 1841). The design for the viaduct consisted of three latticed girders crossing over the main river channel, two of 141 ft and one of 267 ft between bearings, and 15 semicircular masonry arch spans, twelve on the south side and three on the north side of the river, each of 60 ft clear span. The massive masonry piers are composed mainly of local limestone and are founded on rock (except for pier 14).

Initially, the contractor was William Evans of Cambridge, who had raised the tubular girders at the Conwy Bridge on the Chester–Holyhead Railway. However, he went bankrupt on the Boyne Viaduct contract following difficulties in finding a firm foundation for Pier 14. The work was completed in 1855 by the railway company

Test of Boyne
Bridge, 1855

I. ROBINSON, DUBLIN

under the direction of Barton and a young resident engineer, Bindon Blood Stoney (later to become Chief Engineer at Dublin port).

Increasing axle loads led in 1932 to the replacement of the latticed girders with simply supported mild steel girders, the central span being given a curved top chord to improve its appearance. To reduce the maximum load on the bridge, the rail tracks were interlaced, thus allowing only one train at a time across the bridge.

Designed by George Howden, Chief Engineer of the Great Northern Railway of Ireland, the new structure was erected by Motherwell Bridge and Engineering. The viaduct is in constant use conveying intercity passenger and freight traffic between Dublin and Belfast.

HOWDEN G. B. Reconstruction of the Boyne Viaduct, Drogheda. *Trans. Instn Civ. Engrs Ir.*, 1934, **60**, 71–111.

15. Railways across Bogs

HEW 3262
N 500 500

The Midland Great Western Railway Company (MGWR), having acquired the Royal Canal Company in 1845, laid out the first section of their main line from Dublin to Mullingar along the bank of the canal. Between Enfield and Mullingar, however, the canal takes a more circuitous route, having been built to avoid areas of deep bog, in places up to 70 ft deep at the time.

In order to minimize the length of railway, the Engineer to the MGWR, George Willoughby Hemans, convinced that he could find answers to the problem of unstable ground conditions, bravely made the decision to cut straight across the bogs, a total distance of some 8 miles. This involved dewatering the bogs, causing them to consolidate over time. Previous experience of building railways across deep bogs had been limited to Chat Moss in Lancashire, where severe problems were encountered in completing the line of the railway from Liverpool to Manchester.

Essentially, the construction technique involved dewatering the top layers of the bogs by excavating a system of horizontal and lateral drains at appropriate levels across a width of about 75 ft either side of the proposed centre line of the permanent way (see page 58).

The canal in this area is level for about 20 miles and the canal bank blocked the natural drainage of the bogs northwards to the Deal river. Culverts had to be constructed under the canal to lead the water away from the bog drains. One of these culverts was about 120 ft long and 3 ft in diameter and constructed of pine planks held together with iron hoops.

Two courses of heather sods were laid to a width of 30 ft on the consolidated bog surface, which had previously been given a profile rising to 18 in. at the centre, much in the manner of crossfall on a road.

The rails were bolted directly on to longitudinal timbers, which in turn formed part of 25 ft wide lattice frameworks bearing on the prepared bog surface. The result was that no part of the line could deflect suddenly without fracturing, and both sides supported and counterbalanced each other.

Nevertheless, the undulations in the bog surface, when a train passed, were for a while very noticeable, but measurements by the Railway Inspector indicated that deflections did not exceed about 2 inches, so the line was passed and was opened to the public in October 1848.

HEMANS G. W. Account of the construction of the Midland Great Western Railway of Ireland, over a tract of bogs, in the counties of Meath and Westmeath. *Trans. Instn Civ. Engrs Ir.*, 1849–50, **4**, Pt1, 48–60.

RENNISON R. W. (ed.) *Civil engineering heritage: Northern England.* Thomas Telford, London (2nd edn), 1996, 253–4.

R. C. COX

Athlone Rail
Bridge

16. Athlone Rail Bridge

HEW 3079
N 036 420

The Midland Great Western Railway Company (MGWR) was the first to reach Athlone on the banks of the River Shannon. In order to proceed to Galway, then being considered as a possible transatlantic port, George Willoughby Hemans designed a substantial bridge to carry the railway across the river.

The bridge, completed by Fox Henderson & Company in 1851, is formed from twin wrought-iron latticed bowstring girders, 20 ft 4 in. deep at their centres and spanning 176 ft, a central cantilevered opening span of 117 ft 10 in. and two side spans, one of 45 ft 1 in. on the east and one of 54 ft 7 in. on the west side of the river. The girders are carried on piers consisting of pairs of 10 ft diameter cast-iron cylinders spaced 27 ft 11 in. apart and cross braced. The overall length of the bridge is a little over 580 ft.

Compressed air was used when sinking the cylinders to form the piers, the first time in Ireland that this technique of excluding water from the workings was employed.

The side spans of the bridge were renewed in steel

troughing in 1927 and the opening span was fixed in position in 1972. The track was originally double across the bridge, but was later converted to a single line.

SHEPHERD W. E. *The Midland Great Western Railway*. Midland Publishing Ltd, Leicester, 1994, 16 ff.

17. Navan Viaduct

The Navan rail viaduct was constructed in 1850 for the Dublin and Belfast Junction Railway Company by Moore Bros to the design of Sir John Macneill. It carries the 17 mile branch line from Drogheda to Navan over the River Boyne. It consists of seven semicircular masonry arches of varying span crossing the river and two public roadways, the one nearest to the river having been built in more recent times on the flood plain. The spans from the Navan end are three of 49 ft, one of 52 ft, followed by three of 50 ft.

HEW 3133
N 872 675

The viaduct, in coursed rusticated limestone, has an overall length of 406 ft and carries a single track within its width of 27 ft, but accommodates part of the station track layout at the Navan end.

Navan Rail
Viaduct

R. C. COX

18. Obelisk Bridge

HEW 3010
O 047 758

The present bridge was erected in 1868 to replace a timber bridge swept away by floods during the previous year. The bridge spans the River Boyne about 2 miles upstream from Drogheda at a point near to where an obelisk was erected to commemorate the Battle of the Boyne, when in 1690 the Williamite forces defeated the forces of King James. It provides access to the Oldbridge estate from the Drogheda to Slane road.

The bridge is made up of two wrought-iron double-latticed girders, each 128 ft in length, bearing on four 4 in. diameter 2 ft long expansion rollers, which in turn rest on bearing plates on the masonry abutments. The depth between the flanges of the girders is 10 ft 8 in., or one-twelfth the span. The existing abutment on the south side was used, but the northern abutment was extended 28 ft into the river in order to reduce the clear span to 120 ft. The width between the top flanges of the girders is 16 ft. The roadway is supported by buckled plates carried on shallow cross girders at 3 ft 6 in. centres spanning between the main girders.

Obelisk Bridge

The prefabricated girders, each weighing over 28 tons,

R. C. COX

were transported from the works of Thomas Grendon & Co. in Drogheda upriver to the site on a pair of pontoons, and positioned on the abutments at high tide.

Responsibility for the design of Obelisk Bridge was shared by the county surveyors of Louth and Meath (John Neville and Samuel Searanke) and the firm of civil engineers A. Tate. The bridge was designed to carry a distributed load of 350 tons.

STRYPE W. G. Description of the iron lattice girder road bridge, recently erected over the River Boyne, at the Obelisk. *Trans. Instn Civ. Engrs Ir.*, 1871, **9**, 67–78.

19. Dundalk Bay Pile Light

The name 'pile light' refers to the type of construction whereby a lighthouse or other navigation aid is built on a platform supported by a group of piles carried into a sandy sea-bed. The screw pile was invented by a blind Belfast engineer, Alexander Mitchell, and patented by him in 1833. It was used with some success at Dundalk and Cobh, and at a number of other locations in Ireland and abroad. Screw piles are cast-iron columns, usually from 6 to 12 inches in diameter, with screw shaped wings at the toe, so that they would penetrate into the ground by rotation, using a capstan.

HEW 3005
J 119 043

Situated in Dundalk Bay is a pile light marking the entrance to the shipping channel leading to the quays at Dundalk. This light was completed in 1849, but concerns for the stability of the structure on the shifting sand banks in the bay caused a delay in commissioning the light until 1855.

Apart from that at Dundalk, examples of pile lights may be seen on the Spit Bank off Cobh in Cork Harbour (W 811 658), completed in 1852, and in Waterford Harbour off Passage East (S 104 706).

The usual form of construction of such pile lights consists of an octagonal braced structure of 20 ft side at the base, tapering to around 12 ft at platform level. They are of varying height above the sea-bed, but typically around 30 ft. These open iron structures are provided with both lateral and diagonal wrought-iron bracing. Castings were obtained from local foundries.

Dundalk Bay Pile
Light

COMMISSIONERS OF IRISH LIGHTS

LONG B. *Bright light, white water*. New Island Books, Dublin, 1993, 40.

BEAVER P. *A history of lighthouses*. Peter Davies, London, 1971, 58.

MITCHELL A. Submarine foundations, particularly the screw pile and moorings. *Min. Proc. Instn Civ. Engrs*, 1848, **7**, 108–146.

20. Kish Bank Lighthouse

HEW 3050
O 392 304

At the eastern approach to Dublin Bay lies the Kish Bank, a hazard to shipping first marked in 1811 by a light vessel stationed at the northern end of the bank. In 1842, Alexander Mitchell attempted to erect a lighthouse on a platform supported by his patent piles, but bad weather caused the attempt to be abandoned. The bank continued to be marked by a light vessel until the 1960s when the Commissioners of Irish Lights decided to build the present structure.

The resulting lighthouse is a telescopic reinforced concrete structure, comprising an outer caisson of three concentric cylinders of varying height standing on a 3 ft thick base slab and interlocked by twelve radial walls. Outside the centre cylinder, which contains the tower, the caisson is sealed by roof slabs. The diameter of the outer caisson is 104 ft, that of the inner caisson being 41 ft. The total height of the tower from the basement to the helicopter platform is 112 ft. The lower 43 ft of the tower is 35 ft 6 in. in diameter, that of the upper 57 ft 6 in. being 17 ft 6 in.

Kish Bank
Lighthouse

COMMISSIONERS OF IRISH LIGHTS

The tower was constructed inside a caisson at Traders Wharf in Dun Laoghaire Harbour, floated out to the site and sunk onto a prepared bed. The telescopic tower was then jacked up to full height by allowing water to enter the caisson. The water was then pumped out, and the space filled with sand and sealed with concrete.

The lighthouse became operational on 9 November 1965 and was provided with the usual accommodation and operational facilities. The light is now operated automatically and is only visited occasionally for maintenance purposes.

The Kish Bank Lighthouse was designed by R. Gellerstad and built by Christiani and Nielsen of London and Copenhagen. The only other lighthouse of its type in either Britain or Ireland is the Royal Sovereign Lighthouse off Eastbourne (HEW 734).

HANSEN F. Design and construction of Kish Bank lighthouse. *Trans. Instn Civ. Engrs Ir.*, 1965–66, **92**, 247–299.

OTTER R. A. *Civil Engineering Heritage: Southern England.* Thomas Telford Publishing, London, 1994, 215–216.

21. Goodbody's Chimney, Clara

HEW 3254
N 254 330

Dominating the skyline in the town of Clara in County Offaly is the tall chimney at the works of Synthetic Packaging Ltd. The Goodbodys, a Quaker family, established jute mills here in 1865.

The chimney was completed in January 1884 to the design of Robert Goodbody, who also directed its building. It was built of concrete, reinforced with parts of old spinning frames from the mill.

It was originally 150 ft in height, but was reduced to its present height of around 130 ft after being struck by lightning in 1962.

The chimney was a landmark in family history. Lydia Goodbody recalled how, when the full height had been reached, Richard Goodbody, his wife and five children were hauled up the chimney inside in a basket, and, according to family legend, had a picnic on the top!

22. Bohernabreena Reservoirs

In 1880, the Commissioners for the Rathmines and Rath- **HEW 3210**
gar townships approached Dublin Corporation with a **O 092 220**
request to be connected to the Vartry supply. This was to
replace or supplement their existing supply of 2 million
gallons a day, which at that time was being taken from
the Grand Canal at the Eighth Lock near Clondalkin.
Following the failure to agree terms, the Commissioners
went ahead in 1883 with their own scheme to abstract
water from the River Dodder above Bohernabreena in the
Dublin Mountains, using the separation system.

Two impounding reservoirs in the Glenasmole Valley
were formed by the construction of earth dams. The
upper dam is about 600 ft long and reaches a maximum
height above ground level of 70 ft. The lower dam is of
similar length and about 55 ft high. The upper part of the
catchment is an area of granite overlain by peat deposits.
The resulting run-off is highly coloured and is inter-
cepted and led around the upper reservoir in a bypass
canal to the lower reservoir. This part of the supply was
used originally as compensation water for the many mill
owners along the lower reaches of the river, but the total
supply is now piped to the treatment works at Bally-
boden. The larger upper reservoir receives its run-off
from the remainder of the catchment area, which yields
a purer water supply. The reservoir capacities are 360
million gallons and 160 million gallons respectively.

The design, by Richard Hassard, was based on a
scheme put forward by Parke Neville in 1854. The contractors were Falkiner and Stanford and the reservoirs
were in operation by 1887.

ANON. The Rathmines township waterworks. *Irish Builder*, 1884, **26**, 15
Sept., 277.

TYRELL A. W. N. Description of the Rathmines and Rathgar township
water works. *Proc. Instn Mech. Engrs*, 1888, Oct., 523–525.

23. Trim Water Tower

At Trim, County Meath is probably the oldest reinforced **HEW 3257**
concrete water tower remaining in Ireland, and the oldest **N 817 558**
in either Britain or Ireland still in use for water supply.

It was designed for the urban district council by

Trim Water
Tower

R.C. COX

Mouchel & Partners of London and was illustrated in
1909 in *Hennebique Ferro Concrete*. The contractor was
the Irish Hennebique licensee, J. and R. Thompson of
Belfast.

The cylindrical tank has an internal diameter of 30 ft 6
in. and a height of 12 ft. The wall thickness is 14 in. The
base of the tank is supported on a grillage of beams of
four different sizes, carried on six legs.

The 15 in. square legs each have a 7 ft square pad
foundation, all six pads being interconnected by a 'wheel'
of ground beams. At 10 ft vertical intervals, the 30 ft long
legs are braced by a ring beam 12 in. by 9 in. in section.

Inside, the legs are cross braced in a triangle, a different set of legs being so braced at each level.

Although the tank shows signs of slight weeping, a common problem with reinforced concrete water towers, the Trim tower appears to be in remarkably good order.

24. West Link Bridge

The West Link Toll Bridge was constructed between 1988 and 1990 and represents the first large-scale use in Ireland of the segmental balanced cantilever method of construction, whereby the bridge superstructure is built out equally in opposite directions from the previously constructed supporting piers.

HEW 3219
O 077 359

The post-tensioned prestressed concrete viaduct carries the West Link motorway over the valley of the River Liffey near Palmerston to the west of Dublin city. The valley at the site of the bridge is about 550 m wide and 45 m deep. The five spans are 68, 75, 84, 90 and 70 m. The concrete box girders have a constant depth of section of 4 m. There is rigid connection between the deck and the tops of the piers. Expansion joints and sliding bearings are provided at the abutments. The deck slab is 16 m wide and is cantilevered 3.6 m on either side. The piers are of 3 m by 6.5 m section with wall thickness of 750 mm.

Design was by Ove Arup & Partners and the contractors were a consortium of Irishenco and Dycherhoff and Widman, A.G. The West Link Bridge was opened on 12 March 1990.

COLEMAN M. *et al.* West-Link Bridge. *Trans. Instn Engrs Ir.*, 1990–91, **115**, 324–334.

25. Skerries Windmill

The recently restored large windmill at Skerries in County Fingal to the north of Dublin, forms part of a water and wind power complex located in the centre of a town park. There are records of mills on the site as far back as the middle of the sixteenth century. Known as 'The Great Windmill of Skerries', the present four and a half storey tower mill is a reconstruction of the mill which suffered a disastrous fire during a gale about 1860.

HEW 3264
O 250 605

M.LYNCH

The Great
Skerries Windmill

The base of the tower is 20 ft in diameter, reducing to 17 ft 1 in. at the base of the cap. The walls reduce in thickness from 2 ft 8 in. thick at the base to 2 ft 1 in. at cap level about 34 ft above ground level. The overall height to the top of boat-shaped cap is 49 ft. The weight of the cap, including the sails, is estimated to be around 17 tons.

There are five sails, each 33 ft in length, the system of sails being of a primitive spring type consisting of 20 louvres to each arm, connected together, spring loaded and adjusted by hand. The cap has a tail pole and is manually rotated into the wind. The wind shaft is inclined at 16° to the horizontal. The mill ceased operation as a corn mill around the beginning of this century and was restored in 1995–97 by Fingal County Council in association with the Training and Employment Authority and the Skerries Community Development Association.

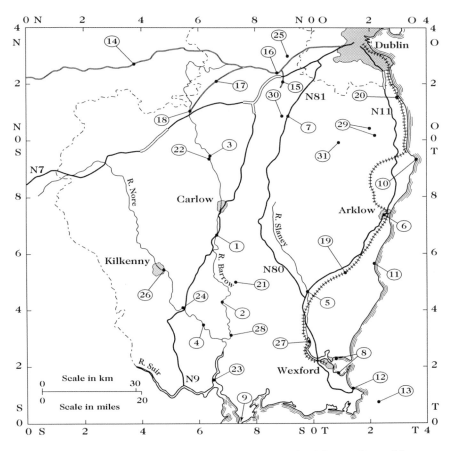

1. Leighlin Bridge
2. Graiguenamanagh Bridge
3. Athy Bridge
4. Inistioge Bridge
5. Scarawalsh Bridge
6. Arklow Bridge
7. Pollaphuca Bridge
8. Wexford Harbour Improvement
9. Hook Head Lighthouse
10. Wicklow Head Towers
11. Courtown Harbour
12. Rosslare Harbour
13. Tuskar Rock Lighthouse
14. Grand Canal (Main Line)
15. Kildare Canal
16. Leinster Aqueduct

17. Grand Canal (Barrow Line and Barrow Navigation)
18. Barrow Aqueduct
19. Dublin and South Eastern Railway
20. Bray Head Tunnels
21. Borris Viaduct
22. Athy Rail Bridge
23. Barrow Rail Bridge
24. Thomastown Viaduct
25. Straffan Suspension Bridges
26. St John's Bridge, Kilkenny
27. 'The Deeps' Bridge, Killurin
28. Mountgarret Bridge
29. Vartry Reservoirs
30. Dublin Water Supply (Blessington)
31. Turlough Hill Pumped Storage Station

3. South Leinster (except Dublin City and District)

Extending southwards from Dublin to Hook Head, and inland as far as the Slieve Bloom Mountains on the borders of County Tipperary, the area is dominated by the major river systems of the Slaney, Barrow, and Nore.

The Dublin and South Eastern Railway reached Wexford by 1872. This difficult route follows the coast as far as Wicklow and has the greatest number of tunnels of any railway in Ireland, notably those at Bray Head. The line was extended to Rosslare Harbour in 1882 and a major development of the harbour facilities was completed there in 1906. These works were undertaken by the railway companies on both sides of the Irish Sea and were aimed at providing a new cross-channel route to Wales. The new line of railway from Rosslare Harbour to Waterford necessitated the construction of a large multi-span steel girder bridge across the River Barrow, east of Waterford.

The inland railway route from Dublin to Waterford, via Carlow and Kilkenny, is carried across the Nore at Thomastown on a viaduct built in 1877.

Apart from Rosslare, there are harbours at Wicklow, Arklow, Courtown, Wexford and Kilmore Quay. Wexford Harbour was the subject of an ambitious improvement scheme begun in the 1840s, mud flats on the north and south shores of the harbour being reclaimed.

There has been a lighthouse on Hook Head since the twelfth century. That on the Tuskar Rock is of more recent origin.

There are many interesting river bridges, notably those over the Barrow. Early examples of the use of reinforced concrete for bridge construction can be found at Kilkenny, Mountgarret, Killurin and Athy.

The main line of the Grand Canal passes through County Kildare, the Barrow line branching off near Lowtown and crossing the Barrow on an aqueduct near Monasterevan. The Barrow Navigation extends from the end of the canal at Athy to the tidal reaches of the river at St Mullins. The

R. C. COX

Leighlin Bridge

Barrow, together with the Nore and the Suir, all flow out into Waterford Harbour and have long provided access to their hinterlands.

Ireland's only pumped storage electricity generation scheme, and an early example of a large impounded reservoir for water supply, are both located in the Wicklow Mountains, the greatest expanse of granite rock on the island.

1. Leighlin Bridge

It is recorded that the crossing of the River Barrow at Leighlinbridge in County Carlow has been important since the tenth century. It is generally accepted that the first stone bridge to be built at this site was that erected by Maurice Jakis, a canon of Kildare Cathedral, in or about 1320. It seems likely that the bridge was rebuilt in the mid-seventeenth century, and widened in 1789.

HEW 3142
S 691 654

The present bridge consists of seven segmental masonry arches of varying span and rise, the largest being just under 30 ft. The rise to span ratio is around 0.3 and the span to pier thicknesses are in the ratio of 5 to 1.

In 1976 a cantilevered reinforced concrete pathway was constructed on the upstream face and the whole bridge pressure grouted and the arch soffits gunited. A bypass was opened in 1986, thus relieving the old bridge of the vibrations and heavy loads imposed by modern commercial traffic.

O'KEEFFE P. J. and SIMINGTON T. A. *Irish stone bridges: history and heritage.* Irish Academic Press, Dublin, 1991, 206–211.

2. Graiguenamanagh Bridge

One of a number of aesthetically very pleasing bridges spanning the River Barrow, the present multi-span masonry arch bridge at Graiguenamanagh was erected in the 1760s to replace an earlier timber structure. Designed

HEW 3136
S 708 438

Graiguenamanagh Bridge

R. C. COX

by George Smith, a pupil of George Semple, the bridge is in the Palladian style.

The spandrels reduce in size towards the ends of the bridge. The arch spans vary from 19 ft 4 in. nearest the river bank to 31 ft 10 in. at the centre of the bridge, the shape of the arches being segmental with a rise to span ratio of 0.4. The arch rings are rusticated using local slaty stone. The piers vary between 7 ft 2 in. and 5 ft 10 in. in thickness. The internal construction of the spandrels is similar to Labelye's work at Westminster Bridge in London.

3. Athy Bridge

HEW 3139
S 681 940

There has been a bridge at Athy since the early fifteenth century, when it was associated with a nearby fortress. A plaque, mounted in a pediment in the parapet over the crown of the central arch on the downstream side, records the fact that the present bridge over the River Barrow at Athy was built in 1796 by Sir James Delehunty, Knight of the Trowel (i.e. a master mason). The bridge is

Athy Bridge

R. C. COX

known locally as Croomaboo, the named coming from 'Crom-a-Boo', the war cry of the Desmonds.

This Palladian style bridge, sponsored by Robert, Duke of Leinster, has five segmental arches, the central span being about 36 ft with a rise of 7 ft 10 in. The width is a generous 30 ft between parapets and there are flanking footpaths. The design was most likely by the engineers of the Barrow Navigation, who carried out works in the area up until 1812.

O'KEEFFE P. J. and SIMINGTON T. A. *Irish stone bridges: history and heritage.* Irish Academic Press, Dublin, 1991, 281–283.

4. Inistioge Bridge

Spanning the River Nore at Inistioge in County Kilkenny, the present bridge represents a rebuilding of an earlier structure, which was severely damaged in the great flood of 1763. There are nine semicircular arches of 24 ft 3 in. span and the width between the parapets is 16 ft 9 in.

HEW 3165
S 636 377

The rebuilding was carried out under the direction of George Smith, the architect responsible for Essex (now Grattan) Bridge over the Liffey in Dublin. Whilst Semple based his Essex Bridge design on Labelye's Westminster

Inistioge Bridge

R. C. COX

Bridge in London, Smith applied the concept of Mylne's design for Blackfriars Bridge to the downstream face of the bridge at Inistioge. It is actually more true to the triumphal bridge model, having all nine arches equal and semicircular, resulting in a parapet which is horizontal rather than curved. The spandrels are of good dark-coloured rubble decorated with pairs of Ionic pilasters in a pale and sharp-edged granite.

RUDDOCK T. *Arch bridges and their builders 1785–1835.* Cambridge University Press, Cambridge, 1979, 105.

5. Scarawalsh Bridge

HEW 3099
S 983 451

The River Slaney rises on the western flanks of Lugnaquillia, the highest peak in the Wicklow Mountains, and flows south before turning south-east to pass through the gap between the Wicklow and Blackstairs Mountains to enter the sea at Wexford. There are a number of masonry bridges spanning the County Wexford reaches of the river, the most important of these being at Enniscorthy and that about 4 miles north of the town at Scarawalsh. This latter bridge carried the main road from Dublin to Wexford over the river until 1976, when a new prestressed concrete bridge was built a short distance

Scarawalsh Bridge

R. C. COX

downstream. An earlier timber structure was carried away in the floods of 1787 and the present masonry bridge was completed in 1790 by the Oriel Brothers from England. The six segmental arches increase in span from 21 ft 6 in. near the banks to 32 ft 6 in. over the main waterway, the overall length of the bridge being about 310 ft. The rise to span ratio is generally 0.25 and there are pointed cutwaters. Scarawalsh Bridge is a good example of a well-built medium sized eighteenth-century masonry bridge, of which there are many other examples throughout Ireland.

6. Arklow Bridge

The main road between Dublin and Wexford passes through the town of Arklow. The town has expanded in recent years and presents a serious bottleneck to through traffic on what is now designated Euroroute E01 from Larne in County Antrim to Rosslare.

HEW 3130
T 247 735

The present route crosses the Avoca River immediately to the north of the town centre on a multi-span bridge erected in 1754–56 by Andrew Noble.

The bridge has a total of nineteen segmental arches, ranging in span from 15 ft to 23 ft, with a rise to span ratio of around 0.18. The original width was about 21 ft 6 in. overall, but this was increased to 37 ft 4 in. when the bridge was widened in 1959. The widening, on the upstream face only, was of reinforced concrete slab construction carried on piles. The earlier masonry work was rehabilitated in 1982. The overall length of the bridge is 495 ft.

Arklow Bridge is soon to be bypassed as part of the upgrading of this strategic route, including a major new bridge river crossing upstream of the old bridge.

7. Pollaphuca Bridge

The design of this spectacular single-span pointed segmental arch bridge is generally attributed to the Scottish civil engineer Alexander Nimmo. It carries the Dublin to Blessington road over the River Liffey at Pollaphuca on the Kildare–Wicklow border. The arch springs from

HEW 3125
N 957 085

Pollaphuca Bridge

R. C. COX

rocks on either side of a gorge through which the river flows out of the Blessington Lakes (artificially created in the 1940s for a hydroelectric scheme). Over the arch, which is ornamented with a bold moulding, runs a cornice, above which is a solid parapet. There are tower abutments with blind and cross-loops and with battlemented parapets surmounting a richly recessed blocking. The spandrels have Gothic niches with hood-mouldings. The arch span is 65 ft, with a rise of 39 ft 4 in., and the maximum height of the bridge above the river is 150 ft.

Pollaphuca Bridge takes its name from the 'puca's pool', a deep pool in the river below a waterfall where legend has it that the puca (a mythical Irish spirit) lived. Despite, or maybe because of, the presence of the puca, Nimmo completed his masterly piece of bridge building in 1820.

O'KEEFFE P. J. and SIMINGTON T. A. *Irish stone bridges: history and heritage.* Irish Academic Press, Dublin, 1991, 274–277.

8. Wexford Harbour Improvement

In 1846, the Wexford Harbour Improvement Company obtained an Act for constructing five embankments, with a view to enclosing much of the area of mud banks or slobs in the harbour. The main objective was to confine the River Slaney to a narrower channel, so that increased scour, assisted by dredging, might create a permanent deep water approach to the quays at Wexford. The chief promoter of the scheme was J. E. Redmond, MP, assisted by William Dargan and others. The company appointed James B. Farrell to be its Engineer.

HEW 3067
T 065 240 to
T 095 240
(north) and
T 060 184 to
T 079 171
(south)

Work commenced on enclosing the north slobs by constructing a clay and marl impounding embankment from the public road at Ardcavan, 1600 yd in a straight line to Big Island, thence curving slightly to the south a further 1830 yd to Raven Point. The crest of this embankment is 6 ft wide and rises 13 ft 6 in. above low water ordinary spring tides. The seaward slope is pitched with stone at 1 in 1½, the landward slope being 1 in 1. An 18 ft wide roadway and a 4 ft wide drain were provided on the landward side. A pumping station was erected south of Ardcavan church at T 078 238, power to the pumps being supplied first by a steam engine, and latterly by electricity. Water is pumped from the Bergerin and Curracloe catchment drains into the sea at low tide through sluices in the embankment.

Having completed the north slobs enclosure works, a further Act was obtained in 1852 to carry out similar, but less extensive works to enclose slobs on the south side of the harbour, a pumping station being erected at Drinagh (T 062 180). The Drinagh station is now derelict, but the pumping station building at the north slobs has recently

R. C. COX

Outlet sluices and pumping station (north slobs, Wexford Harbour)

been restored by the Office of Public Works. Parts of the north slobs have been occupied by the Wexford Wildfowl Reserve since 1969 and are a wintering ground for about half the European population of Greenland white-fronted geese.

ANDERSON W. On the reclaimed lands at Wexford harbour, and the machinery employed in draining them. *Trans. Instn Civ. Engrs Ir.*, 1862, 7, 102–123.

9. Hook Head Lighthouse

HEW 3068
X 732 972

Generally regarded as the site of one of the oldest light-houses in either Britain or Ireland, the present Hook Head Lighthouse on a headland in the south-west of County Wexford began as a fortress and lookout tower. It was built by the Norman, Raymond Le Gros, in 1172 at the entrance to Waterford Harbour. Welsh monks are reputed to have maintained a crude form of warning beacon here from as early as the fifth century.

In the thirteenth century, William Marshall established a light on the tower to guide ships up the estuary to the new port of Ross (now New Ross). The light was an open fire maintained by local monks from nearby Churchtown.

The tower was altered and enlarged over the years, the present lantern and supporting turret being erected in 1864 to replace a smaller lantern and turret of 1792.

The tower has three vaulted chambers and a spiral staircase. The walls of the tower vary in thickness from 12 ft 10 in. to 9 ft. The earlier part of the tower is about 75 ft in height (four storeys) with a diameter of about 38 ft. The final 20 ft of the tower to the level of the balcony is about 20 ft in diameter. The overall height to the top of the lantern is 118 ft. The diameter of the main tower is about 40 ft, but narrows somewhat towards the top.

Hook Head
Lighthouse

R. C. COX

Lighting was by oil lamps from 1791 until 1871, when coal gas, provided from a small gasworks, was used. This was replaced by vaporized paraffin in 1911 and then by electricity in 1972. The families of lighthouse keepers lived at the Hook until 1977 when the lighthouse became 'relieving'. It was converted to automatic working in 1996.

LONG B. *Bright light, white water*. New Island Books, Dublin, 1993, 63–67.

HAGUE D. B. and CHRISTIE R. *Lighthouses: their architecture, history and archaeology*. Gomer Press, Llandysul, 1975, 14 ff.

10. Wicklow Head Towers

HEW 3155
T 345 923

In an effort to light one of the main hazards to shipping along the east coast, the Engineer to the Revenue Commissioners, John Trail, in 1781 built twin octagonal towers on Wicklow Head, one of the headlands along the east coast requiring navigation marks. The towers were built of cut stone with dressed granite quoins and window detailing. However, a common problem with these early towers was their height above sea level. Many were built too high up the cliffs and, as a result, were often obscured by fog, as was the case here.

In 1818, a new upper tower, 78 ft high, was erected, under the supervision of George Halpin (Senior), on the site of the earlier lower tower, but use of this upper tower was discontinued in 1865 and the lantern removed in 1978. A new front tower, 45 ft high, was built much lower down the cliff and is still in use. The present lantern, blocking balcony and railing were added in 1868.

The original lower, or front, tower on the saddle of the headland was demolished in 1818. The upper, or rear, tower on Long Hill was struck by lightning in 1836 and the ensuing fire gutted the inside of the tower. In order to protect the tower from the elements and to continue its use as an additional navigation mark, a brick dome was added in 1866. The tower is still extant and has recently been rehabilitated by the Irish Landmark Trust for use as as a luxury holiday home. It is 88 ft 8 in. in height to base of parapet level, the 9 ft 3 in. diameter dome adding a further 8 ft 10 in. to the overall height. The tower is

29 ft 4 in. across at the base, tapering to 20 ft 11 in. across
at the top.

LONG B. *Bright light, white water.* New Island Books, Dublin, 1993, 48–52.

11. Courtown Harbour

Courtown Harbour lies on the Wexford coastline east of
Gorey, approximately midway between Kilmichael
Point to the north and Cahore Point to the south. The
prime promoter of the harbour here was Lord Courtown,
who in 1824 obtained an Act of Parliament (5 Geo.IV
c.122) granting the necessary finance to the Courtown
Harbour Commissioners to enable design and construction to begin.

HEW 3037
T 202 563

The contractors O'Hara, Simpson and McGill commenced work on a pier to the south at Breanoge Point and
followed a north-east direction as previously suggested
by the government engineer Alexander Nimmo in 1819.
Asked for his opinion in 1825, George Halpin (Senior),
the engineer to the Ballast Board at Dublin, foresaw problems of siltation. In 1831, Sir John Rennie's opinion was
sought and by 1833 Owen of the Board of Works, in
whose charge the harbour had been placed, reported that
the pier was breaking up.

Courtown
Harbour

R.C. COX

The decision was then taken to build a new south pier at right angles to the shore near the mouth of the Breanoge river and to develop the harbour to more or less its present form. Under the direction of Francis Giles, in 1834–47, the south pier, followed by a new north pier, were completed, the harbour was deepened, and quays built. Sluice gates and an overfall were provided, and a system was designed to maintain the water level in the harbour at low tide and to provide a means of scouring the narrow entrance channel between the piers. The harbour was substantially completed by 1847. The south pier was rebuilt and lengthened in 1871, when Isaac Mann carried out repairs to the harbour. New sluice/tidal gates were provided in 1891 and in 1905 the responsibility for the harbour was transferred to Wexford County Council, who carried out a partial rebuilding and lengthening of both piers in the 1960s.

KINSELLA A. *The windswept shore: a history of the Courtown district.* Anna Kinsella, Dublin, 1982, 55–71.

12. Rosslare Harbour

HEW 3092
T 136 126

Rosslare Harbour is at the south-east corner of County Wexford near Greenore Point, and about 8 miles south-east of Wexford town. In 1873, the Rosslare Harbour Commissioners built a 480 ft long timber jetty to provide two berths for steamers from Fishguard in south Wales. This was connected to the shore by a 1000 ft long viaduct.

When the Dublin, Wicklow and Wexford Railway was extended from Wexford to Rosslare in 1882, a new jetty and viaduct were provided, the jetty being dog-legged and mostly of iron and timber construction.

Following the setting up of the Fishguard and Rosslare Railways and Harbours Company by the Great Western Railway in England and the Great Southern and Western Railway, a new harbour was built between 1904 and 1906. This consisted of a double-track railway viaduct of eleven 81 ft spans of steel girders on cast-iron cylindrical piers. This led to a new jetty, 1550 ft in length. New quay and sea protection walls were also completed in mass concrete and land reclaimed from the sea using sand obtained from suction dredging of the approaches to the harbour.

The main contractor for the 1882 works was W. Murphy of Dublin and for the 1906 contract, Charles Brand & Company of Glasgow. Only traces of the works described above now remain as the harbour was completely rebuilt and enlarged, first in the late 1960s and more recently in the 1990s. Rosslare Harbour is now a major ferry terminal and container port.

CASSERLEY H. C. *Outline of Irish railway history*. David and Charles, Newton Abbot, 1974, 34.

ANON. Opening of a new route to Ireland. *The Engineer*, 1906, **102**, 7 Sept., 239–242.

13. Tuskar Rock Lighthouse

In 1810, when the Corporation for Preserving and Improving the Port of Dublin took over the management of Irish lighthouses, they were very aware of the dangers presented to shipping by the unmarked and unlit Tuskar Rock. The Tuskar Rock lies some 5 miles south-east of Greenore Point in County Wexford and is directly in the path of all shipping entering or leaving the port of Rosslare.

George Halpin (Senior), the Corporation's engineer, inspected the rock and proposed a design for a lighthouse similar to that previously erected on South Rock off the coast of County Down. The Elder Brethren of Trinity House were also asked for their opinion. Their preference was for a tower similar to that designed by D. A. Alexander and erected at the South Stack near Holyhead. In the event the design was a compromise and work commenced on the rock in 1812.

The Tuskar Rock Lighthouse consists of a tapering bell-bottomed granite tower with an overall height of 111 ft. The tower is 36 ft diameter at the base and 17 ft diameter at balcony level. The lantern, blocking balcony and railing were renewed in 1885. The buildings around the base of the tower were added at various times.

The building of lighthouses is, by its very nature, a hazardous occupation, particularly on wave-swept rocks. Workmen could be besieged for days on rocks by bad weather and continually risked their lives. Surprisingly there were relatively few accidents, but sadly,

HEW 3118
T 226 071

eleven persons lost their lives on the Tuskar Rock during a severe storm at an early stage in the construction of the tower.

The light on the Tuskar Rock was first exhibited on 4 June 1815.

LONG B. *Bright light, white water*. New Island Books, Dublin, 1993, 53–59.

14. Grand Canal (Main Line)

HEW 3001
O 177 345 to
N 033 189

In 1715 an Act was passed in the Irish Parliament 'to encourage the draining and improving of the bogs and unprofitable low grounds and for the easing and despatching the inland carriage and conveyance of goods from one part to another within the Kingdom'. Commissioners for Inland Navigation were appointed in 1729 and funds were made available to promote the aims of the Act. The Act marked the beginning of the canal age in Ireland.

Work started in 1757 to connect Dublin by canal with the Shannon Navigation via the Grand Canal. Thomas Omer surveyed the route and, following the incorporation of the Grand Canal Company in 1772, work was continued by John Trail, and later Charles Tarrant. The first section from the Dublin terminus at City Basin (St James's Street) to Sallins was opened to traffic in 1779; it had taken 23 years to construct 18 miles of canal.

The canal crosses the River Liffey by the Leinster Aqueduct and continues to its summit level near Lowtown in County Kildare, some 280 ft above low water at Dublin. Near Lowtown the canal obtains a water supply from the Milltown and Blackwood feeders.

Omer had planned that the canal would cut straight across the Bog of Allen, but John Smeaton advocated a more northerly route towards Edenderry, which was adopted. Smeaton had argued with William Chapman about the desirability or otherwise of draining the bog prior to cutting the canal. Smeaton's view prevailed and the canal was constructed at the same level as the existing surface of the bog, without allowing for a period of drainage. The result was that the land on either side drained into the canal and subsided, leaving the canal confined by high embankments. These have proved to be

R. C. COX

a constant source of trouble to the canal: major breaches of the banks occurred soon after completion, again in 1916, and more recently in 1975.

Sixteenth Lock, Grand Canal

Richard Evans and his assistant John Killaly progressed slowly with much difficulty, and the line was opened to Daingean by 1797 and to Tullamore in 1798. Tullamore Harbour became the temporary terminus of the Grand Canal whilst it was decided how best to reach the Shannon.

The route eventually chosen followed the valley of the River Brosna, joining the River Shannon just north of Banagher at Shannon Harbour. Richard Griffith (Senior), the father of Sir Richard Griffith, supervised the work with John Killaly as his engineer. The earlier lessons having been learnt the hard way, some 3000 men spread amongst 22 contractors toiled to drain the bog before this final section of the canal was built. The canal was finally opened to the Shannon in 1804. In 1950 the Grand Canal

Company was merged with Córas Iompair Éireann and in 1986, responsibility for inland waterways passed to the Office of Public Works, and latterly to the Department for Arts, Culture and the Gaeltacht.

The average width of the canal is 30 ft and the average depth at the centre is 5 ft with a minimum headroom under the bridges of 8 ft 6 in. at the waterline. There are five aqueducts, and dry docks at Tullamore and Shannon Harbour. Of the 36 locks, 30 are single and six double (to conserve water). The average width of the locks and the navigation under bridges is about 15 ft. The shortest lock is 69 ft 8 in, the longest 88 ft 7 in, the narrowest 13 ft 6 in, and the widest 16 ft 2 in. Branches of the canal were completed to Athy (the Barrow Line), and to Edenderry, Kilbeggan and Ballinasloe. The branch to Naas and onwards to Corbally (known as the Kildare Line) was acquired in 1808 from the County of Kildare Canal Company. The Grand Canal was connected to the River Liffey in 1796 by the construction of the Circular Line from the First Lock at Inchicore to the Grand Canal Dock near Ringsend.

DELANY R. *The Grand Canal of Ireland.* David and Charles, Newton Abbot, 1973.

15. Kildare Canal

HEW 3081
N 883 225 to
N 840 140

In 1789 the County of Kildare Canal Company (CKCC) completed a 2¾ mile canal from a junction with the Grand Canal near Sallins to Naas in County Kildare. The line had earlier been surveyed by Richard Evans, but the canal was constructed under the direction of William Chapman, who built three skewed masonry arched bridges over the waterway (the first of their type in either Britain or Ireland) in order to avoid realigning the road crossings. Chapman came to Ireland as an agent for the steam engine builders Boulton and Watt, and decided to stay on and make a name for himself as a canal engineer.

The branch has five locks, originally built to Chapman's design, and served the Leinster mills at the Second Lock and Naas itself. The main cargoes were flour from the mills and coal to Naas.

When the CKCC went bankrupt, it was purchased by the Grand Canal Company who, in 1808, proceeded to replace the bridges, provide towpaths and enlarge the locks to match those on the main line (73 ft 6 in. by 14 ft by 5 ft deep). A 5¾ mile extension without locks was also constructed as far as Corbally and opened in 1811. The contractors were Henry, Mullins and Mac-Mahon.

A low-level road bridge near Naas, built in the 1950s, has since blocked the navigation upstream. The remainder of the Kildare Canal was closed to navigation in 1961 but it is hoped to carry out full restoration. In the meantime, single lock gates have been provided to maintain the water levels in the canal for aesthetic and maintenance reasons.

DELANY R. *Ireland's inland waterways*. Appletree Press, Belfast, 1992, 83.

16. Leinster Aqueduct

The Leinster Aqueduct carries the Grand Canal over the River Liffey to the west of Sallins in County Kildare.

HEW 3096
N 876 228

The location of the aqueduct was a matter of great debate and argument amongst the engineers of the day. Thomas Omer at first planned to lock down to the river and lock up again on the far side but, following reports from both Smeaton and Jessop in 1773, an aqueduct was built on the site previously selected in 1771 by Charles Vallencey, acting for the Commissioners of Inland Navigation.

The aqueduct was designed by Richard Evans, the Engineer to the Grand Canal Company, and construction work commenced in 1780. The aqueduct, constructed in ashlar limestone masonry, has five arches, each of 25 ft span, with a rise of 7 ft 6 in., and has pointed cutwaters. The overall length is 233 ft, the width of the river at this point being 152 ft. There is a 10 ft 9 in. wide towpath on the north side, a 15 ft 6 in. wide roadway on the south side, and a canal waterway of 16 ft 6 in. width. Including the 18 in. thick parapet walls, the overall width of the aqueduct is 46 ft 3 in. Another aqueduct, with two small arches, lies some 100 ft further to the west of the end of the main aqueduct.

R. C. COX

Leinster Aqueduct   There is provision for excess water to overfall from the canal to the River Liffey below and, in times of water shortage, the process may be reversed using temporary pumping facilities.

DELANY R. *The Grand Canal of Ireland*. David and Charles, Newton Abbot, 1973, 23–23, 29.

17. Grand Canal (Barrow Line) and Barrow Navigation

HEW 3003
N 778 254 to
S 682 935

Between 1783 and 1791, the Grand Canal Company built a branch from the main line near Lowtown to the River Barrow at Athy in County Kildare, a distance of 28 miles. The descent of approximately 100 ft from the summit level at Lowtown required nine locks and a small number of aqueducts – the principal one, of three 40 ft spans, carrying the canal across the River Barrow at Monasterevan in County Kildare. There was also a further branch of 10 miles to Mountmellick via Portarlington, but this is now derelict.

The designers involved were principally Richard Evans and Archibald Millar. The contractors for the section from Lowtown to Monasterevan were Thomas Black and others, and John MacMahon completed the remainder to Athy. At Athy, canal traffic is locked down to the River Barrow.

The Barrow Navigation consists of 42 miles of river navigation, 11 miles of which are artificial cuts or canalized sections, the longest being the Levitstown cut of 2 miles between Athy and Carlow. There are 23 single-chambered locks between Athy and the sea lock at St Mullins at the tidal limit of the river. The locks for the most part are 80 ft long by 16 ft 6 in. wide. A total of 22 weirs along the river maintain water levels upstream of the canalized sections.

HEW 3004
S 682 935 to
S 725 380

Work commenced on improving the Barrow Navigation as early as 1715, but was not completed until 1791.

DELANY V. T. H. and DELANY R. D. *The canals of the south of Ireland.* David and Charles, Newton Abbot, 1966, 30 ff., 126–138.

DELANY R. *Ireland's inland waterways.* Appletree Press, Belfast, 1992, 68–71, 80–82.

Athy Lock

R. C. COX

18. Barrow Aqueduct

HEW 3127
N 622 105

The Grand Canal Company, under the direction of their engineer Richard Evans, had completed the Barrow Line as far as Monasterevan by 1785. This included facilities for locking down to the River Barrow and up the other side to continue the extension of the canal to Athy. This somewhat cumbersome method of passing the canal traffic across the river was replaced in 1831 by the opening of the present aqueduct.

The Barrow Aqueduct was designed by Hamilton Killaly (a son of John Killaly) and built between 1827 and 1831 by Henry, Mullins and MacMahon. It consists of three segmental masonry arches, each of 41 ft 6 in. span, with a rise of 7 ft 6 in., giving a low rise to span ratio of 0.18. The arch voussoirs are rusticated ashlar limestone and the garlanded decoration over each river pier gives the aqueduct an attractive overall appearance.

The canal over the aqueduct is 17 ft wide, with an 8 ft wide towpath on either side. The overall length of the aqueduct is 207 ft, 134 ft 6 in. of this being over the waterway. At the eastern end of the aqueduct is a small lifting bridge and nearby is a group of canal warehouses.

Lifting bridge and canal warehouses on the Grand Canal near Monasterevan

DELANY R. *The Grand Canal of Ireland.* David and Charles, Newton Abbot, 1973, 97–98.

R. C. COX

19. Dublin and South Eastern Railway

The Dublin and South Eastern Railway (D&SER) served the area to the south-east of Dublin, including the counties of Wicklow and Wexford and Waterford city; the history of the railway has been ably recounted by Shepherd. The earlier Dublin, Wicklow and Wexford Railway (DW&WR) became the D&SER in 1907, the latter amalgamating with the Great Southern and Western Railway (GS&WR) in 1925.

HEW 3101
O 167 339 to
T 050 221

The engineering problems of the route from Dublin to Wexford included tunnelling through the headlands at Dalkey and Bray, bridging a number of river valleys and combating the ever present effects of coastal erosion.

The route from Dublin to Bray via the outlying villages of Foxrock and Shankill, known as the Harcourt Street Line, crossed the River Dodder on a fine masonry arch viaduct of nine spans (O 166 301) and the Shanganagh River near Shankill on a smaller viaduct of similar design.

The DW&WR took a lease of the Dublin and Kingstown Railway in 1856 and converted it to the Irish standard gauge of 5 ft 3 in. The alignment of the atmospheric system extension from Kingstown (now Dun Laoghaire) to Dalkey was also used as part of the extension of the railway towards Bray. Between Dun Laoghaire and Sandycove, the track is laid in a deep cutting between vertical retaining walls. This line was continued to Bray along the coast at Killiney, necessitating the driving of a 492 ft long curving tunnel through solid granite at Dalkey and extensive earthworks along the steep sided coastline at Killiney Bay. The consulting engineer to the company from 1845 to 1859 was Isambard Kingdom Brunel. He was on hand to advise the company as to how to overcome the considerable engineering problems of carrying the railway around Bray Head.

Further south, the viaduct over the Avonmore River at Rathdrum and a tunnel under the town of Enniscorthy are worthy of note. The Rathdrum Viaduct, completed in 1861, has five 44 ft semicircular arched spans, the rail level being about 88 ft above the river bed. There are also two tunnels to the south of the viaduct.

The engineering work as far as Wicklow was in the hands of the experienced railway contractor, William

Dargan, and Thomas Edwards was responsible for the construction of the line between Wicklow and Wexford. Engineers appointed to the railway companies in the early years of the east coast route included Barry Gibbons, William Le Fanu, Thomas Grierson and John Chaloner Smith.

SHEPHERD W. E. *The Dublin and South Eastern Railway*. David and Charles, Newton Abbot, 1974.

20. Bray Head Tunnels

HEW 3123
O 285 173 to
O 285 147

The main railway route from Dublin to Wexford and Rosslare Harbour follows the coast for much of the way as far as Wicklow and the engineers and contractors had many problems to overcome in protecting the line from the ravages of the sea. The line around Bray Head, between Bray and Greystones, passes through a series of tunnels and across a number of short bridges and embankments.

Bray Head rail
tunnels (right,
1855; left, 1876)

In 1855, Isambard Kingdom Brunel designed three tunnels for double track, but only single track was laid. Built by William Dargan, the tunnels were the Brabazon (210 yd), Brandy Hole (300 yd) and Cable Rock (143 yd)

R. C. COX

The rocks of Bray Head are Precambrian and very hard, making tunnelling expensive, but they are also very unstable. There is a always a risk of rock falls on to the line and there are daily inspections.

The Brabazon Tunnel was bypassed in 1876 by two short tunnels. Two timber trestle viaducts, to the south of Brandy Hole, were replaced by a single masonry arch and two arched sections built to support the cliff above the track. A further deviation became necessary in 1889, including another bridge replacement, and the Brandy Hole and Cable Rock tunnels were relined with brick and stone without altering the loading gauge.

Further erosion at the base of the cliffs necessitated a final major deviation. The Long Tunnel, designed by C. E. Moore, was commenced in 1913 by Naylor Brothers of Huddersfield and completed in 1917 by direct labour under the resident engineer W. H. Hinde, after Naylor had withdrawn from the contract. The tunnel is 1084 yd long, 15 ft in internal height and has a 9 ft 6 in. diameter ventilating shaft 100 ft deep.

MURRAY K. A. Bray Head. *Journ. Ir. Rlwy Rec. Soc.*, 1980, **14**, no. 82, 71–84.

21. Borris Viaduct

In 1854, some ten years before the opening of the Dublin, Wicklow and Wexford Railway (DW&WR) to Enniscorthy, the Bagenalstown and Wexford Railway was incorporated in a bid to reach Wexford by an inland route. Although the route selected ran through wild and unremunerative countryside, there was some prospect of through traffic.

HEW 3110
S 730 500

By the end of 1858 the standard gauge line had been completed from the junction with the Great Southern and Western Railway at Bagenalstown as far as Borris in County Carlow, where the original station buildings are now used as residences. The line was completed to Ballywilliam by 1862, by which time the company had run out of funds and was declared bankrupt. Links with Wexford (via Macmine Junction) and New Ross (via Palace East Junction) were eventually completed, but by then the DW&WR had reached Wexford.

Passenger services were terminated in 1931, but occa-

R. C. COX

Borris Viaduct

sional freight trains worked the line up until closure in 1963.

To the south of Borris station, the line was carried over the Mountain river valley on a substantial limestone viaduct of 16 spans. The semicircular arches are each of 35 ft. diameter and are carried on slightly tapering piers some 10 to 15 ft above the average ground level. The piers are 5 ft thick and 18 ft wide at the base. The viaduct passes over a public roadway at the town end and has a total length of about 470 ft.

The engineer to the railway company was William Le Fanu and the main contractor for the line was John Bagnall. There is pedestrian access to the viaduct as part of a nature trail.

MURRAY K. A. and MCNEILL D. B. *The Great Southern and Western Railway*. Irish Railway Record Society, Dublin, 1976, 36.

22. Athy Rail Bridge

HEW 3246
S 682 934

During World War I, there was growing concern about the security of fuel supplies and determined efforts were made to work the minor coal deposits in South Leinster. In order to provide rail access to the Wolfhill Collieries, a 10 mile branch line was laid in 1918 by the Great Southern and Western Railway Company from the main line south of Athy in County Kildare. The single-track branch line, now reduced to a short spur to a cement products factory, is carried across the River Barrow on a six-span reinforced concrete bridge.

The bridge comprises a pair of longitudinal beams, 5 ft 9 in. deep by 2 ft thick, spanning 39 ft between supports. A slab carries the permanent way and the parapets are supported on cantilevered beams at 7 ft centres.

ROWLEDGE J. W. P. *A regional history of railways, Volume 16: Ireland.* Atlantic Transport Publishers, Penryn, Cornwall, 1995, 89–90.

23. Barrow Rail Bridge

A new rail route across the south of County Wexford, between Rosslare Harbour and Waterford, necessitated the construction of a number of bridges, viaducts and tunnels. The major construction project was the crossing of the estuary of the River Barrow at its confluence with the River Suir, some 4½ miles west of Campile in County Wexford and opposite Cheek Point in County Waterford.

Designed by Sir Benjamin Baker, who was also in-

HEW 3075
S 683 147

Barrow Rail Bridge

R. C. COX

volved in the design of the famous Forth Rail Bridge in Scotland, the Barrow Bridge is, at 2132 ft, the longest rail bridge spanning entirely across water in Ireland and was, at the time of its completion in 1906, the third longest in either Britain or Ireland. The engineer to the GS&WR at the time was James Otway. At the west end of the bridge the railway enters a tunnel 217 yd long.

The bridge consists of eleven fixed spans of 148 ft each, two end spans of 144 ft each and a centrally pivoted opening span of 215 ft. The spans consist of mild steel Pratt-type latticed girders supported on cast-iron cylinders, varying in diameter from 8 ft to 12 ft, infilled with concrete and founded on the underlying rock. Cross girders spanning between the latticed girders support the single line track. There is also overhead cross bracing between the main girders.

The main contractor for the bridge was Sir William Arrol & Co. of Glasgow, the contractor for the line between Rosslare and Waterford North station being Sir Robert McAlpine & Sons.

ANON. The Fishguard and Rosslare route to Ireland. *The Engineer*, **102**, 1906, 7 Sept., 239–242.

24. Thomastown Viaduct

HEW 3109
S 576 410

With the completion in 1850 of a viaduct over the River Nore to the south of Thomastown, the Great Southern and Western Railway commenced rail services between Dublin and Kilkenny. The main span of the original viaduct was formed of timber latticed girders, and was the longest rail bridge span in either Britain or Ireland at the time. It was designed by W. S. Moorsom and constructed by the firm of J. and R. Mallet of Dublin. The masonry work was by Hammond and Murray of Dublin.

This span was replaced in 1877 by the present structure, designed by C. R. Galwey and supplied by Courtney, Stephens and Bailey of Dublin. It consists of twin wrought-iron bowstring girders of 212 ft span, bearing on cast-iron plates bedded into the rough hewn coursed masonry abutments. The depth of the girders at mid-span is 25 ft 6 in. and 5 ft 9 in. at the ends. The web of each girder consists of a diagonal bracing arranged in a double

M. B. BARRY

system of triangulation, dividing the flanges into 15 equal bays, with stiffening plates at each end. Cross girders are suspended from the bottom flanges at the intersection of the diagonals. Each end of the main span is approached by twin masonry arch viaducts, the central piers of which incorporate transverse relieving arches. The total length of the viaduct is 428 ft and the rail level is 78 ft above river bed level.

Thomastown Viaduct

GALWEY C. R. On the Nore Viaduct at Thomastown. *Trans. Instn Civ. Engrs Ir.*, 1879, **12**, 133–151.

25. Straffan Suspension Bridges

Large country estates frequently included linear water features which necessitated the provision of at least one bridge. The landed gentry were able to afford iron or high quality masonry arch bridges or, in some cases, suspension bridges, as at Birr Castle and as here at Straffan, the former estate of the Barton family in County Kildare.

    Spanning an artificial channel, fed from the River Liffey, this small footbridge was erected in 1849 by the firm of Courtney and Stephens of Dublin. The suspension bridge spans 45 ft and is comprised of twin cast-iron

HEW 3038
N 918 295

columns carrying the suspension cables, the decking being suspended by vertical hangers from the cables. On top of the columns, there are ornamental finials. There is a handrail parallel to the decking, cross braced with diagonal tie bars. The decking was later concreted in order to stiffen the structure for public use.

There is a second bridge of similar construction, although much altered, spanning the main channel of the river nearby.

Straffan
Suspension Bridge

R. C. COX

26. St John's Bridge, Kilkenny

Crossing the River Nore beside the great castle of the
Butlers in Kilkenny, this reinforced concrete arch of 140 ft
span is an early example of the use of what was then
popularly known as ferroconcrete. This single arch replaced a three-span masonry arch bridge erected in the
1760s and designed by William Colles, the founder of the
Kilkenny Marble Works.

HEW 3167
S 508 558

The design of the present bridge was by Mouchel &
Partners of London and Alexander Burden, the County
Surveyor of Kilkenny. The work was carried out by J. and
R. Thompson of Belfast and Dublin, a licensee of the
Hennibique system of reinforcing concrete. The system
of reinforcement consisted of 'a combination of alternate
straight bars and bars with ends bent up at an angle,
with vertical U-bars, or stirrups, of flat iron passing
around the straight bars and reaching nearly to the top of
the beams'.

The arch has three ribs 2 ft wide and spaced 12 ft apart
centre to centre, spanning from concrete buttresses,
which are carried on precast piled foundations. The
thickness of the arch at the crown is 2 ft 6 in. Each rib is
2 ft 6 in. thick at the springings and 2 ft at the crown. The
main longitudinal beams are 14 in. by 24 in. deep. When
built, the bridge had open spandrels consisting of 14 columns of 12 in. square section.

In 1970, it was found that the piles had moved hori-
zontally and it was decided to remove the road surfacing
and place a concrete slab over the decking. At the
same time, the spandrels were filled in, thus radically
altering the way the structure reacts to the loads placed
upon it.

When completed in 1912, St John's Bridge was the
longest span reinforced concrete road bridge in either
Britain or Ireland.

ANON. Reinforced concrete bridge over the Nore, Kilkenny. *Conc. and
Constr. Eng.*, 1911, **6**, no. 3, 223–226.

27. 'The Deeps' Bridge, Killurin

HEW 3179
S 975 268

The River Slaney between Enniscorthy and Ferrycarrig in County Wexford is spanned by an early reinforced concrete road bridge at an area known as 'The Deeps' near Killurin. The present bridge replaced a timber trestle bridge erected in 1842–44 under the direction of the County Surveyor, J. B. Farrell.

The present bridge consists, on the eastern side of the river, of five spans, each of 30 ft, approached by a 216 ft long embankment terminating in a masonry abutment. On the western side of the waterway there are five spans, four of 30 ft and one of 19 ft 6 in. Between these spans there is a bascule-type steel lifting span of 40 ft opening, placed between two fixed sections 10 ft 6 in. long. The supporting piers consist of reinforced concrete pile groups carried up to the deck and cross braced. The overall length of the bridge is 609 ft, 362 ft being over the waterway. The overall width is 20 ft 4 in. and 18 ft 4 in. between parapets.

The bridge was designed by Delap and Waller of Dublin and erected in 1915 by the British Reinforced Concrete Engineering Company, using BRC Fabric as the reinforcement. A contemporary company advertisement

'The Deeps'
Bridge

R.C. COX

described this fabric as 'a continuous wire mesh of drawn steel wires. It is laid in the concrete along the lines where the tension is greatest and each wire takes its share of the work.' The lifting span was supplied by the Cleveland Engineering Company.

ANON. Reinforced concrete roads. *The Irish Builder and Engineer*, 1915, **57**, no. 19, 399.

28. Mountgarret Bridge

The replacement of a timber bridge spanning the River Barrow at Mountgarret, erected by the American Lemuel Cox around 1795, was one of a number of major civil engineering projects completed in the early 1930s following the foundation of the Irish Free State. The bridge carries the Thomastown to New Ross road over the river immediately before its junction with the Enniscorthy to New Ross road about 3½ miles upstream of New Ross.

HEW 3088
S 726 305

It consists of four spans of 29 ft 6 in., one of 23 ft 6 in. forming the platform for an opening span of 47 ft, and two end cantilevered sections, each of 10 ft 6 in. span. The bridge is of reinforced concrete with a steel opening span

Mountgarret
Bridge

R. C. COX

of the Scherzer rolling lift type. The superstructure is carried on 5 ft 3 in. diameter piers, founded about 5 ft below river bed level, and supported on groups of 14 in. by 14 in. reinforced concrete piles driven to rock. The piers either side of the opening span are protected by timber dolphins, the clear navigation opening being 40 ft. The width between parapets is 18 ft and the roadway width 16 ft 6 in. The approach embankments are 140 ft in length on the Wexford side and 450 ft on the Kilkenny side of the river.

The bridge was designed by A. H. Delap of Delap and Waller, the general contractor being John Hearne & Co. The steelwork for the opening span was supplied and erected by John Butler & Co. of Leeds.

MACONCHY J. K. Mountgarrett Bridge. *Trans. Instn Civ. Engrs Ir.*, 1931–32, **58**, 147–231.

29. Vartry Reservoirs

HEW 3019
O 218 018

The first public water supplies for Dublin were drawn from wells, local streams and rivers, and later the canals. In 1860, the Hawkshaw Commission recommended that a gravity supply be provided from a surface reservoir to be constructed near Roundwood in County Wicklow, as originally proposed some years earlier by the engineer-

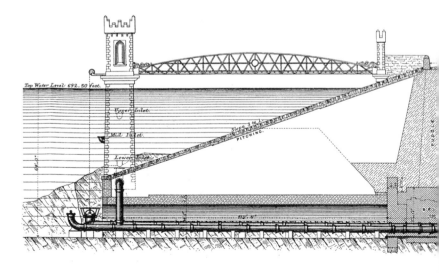

ing consultant Richard Hassard. The Dublin City Engineer, Parke Neville, drew up the plans for the scheme, Thomas Hawksley acting as consultant. The necessary Bill was steered through Parliament in 1861 by Sir John Gray, Chairman of the Waterworks Committee, and the foundation stone was laid at Roundwood in November 1862.

The Lough Vartry impounding dam and reservoir are at an elevation of about 690 ft above sea level and about 8 miles from the source of the River Vartry, which rises in the Wicklow Mountains to the west.

The dam, consisting of a 1640 ft long earthen embankment with puddled clay core, reaches a height of 71 ft. It is 27 ft 10 in. wide at the crest and 377 ft 3 in. at its lowest point. It carries the road from Roundwood to Wicklow along its crest.

The 410 acre impounded reservoir has a capacity of around 25 million gallons and a greatest depth of about 60 ft. The draw-off tunnel under the embankment, which accommodates the supply pipes from the draw-off tower to the treatment works, is 14 ft square with a semicircular arched roof. Access to the intake tower is by a 97 ft span iron girder bridge consisting of twin latticed trusses with parabolic curved top chords.

Roundwood
Dam (sectional
elevation)

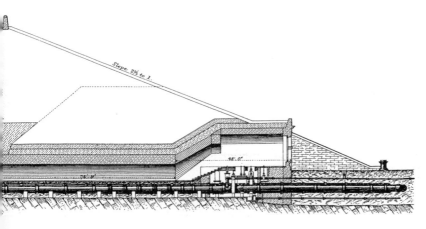

The main contractors for the 1862–67 scheme were William McCormick (London and Derry), all the iron-work being supplied and installed by Edington & Son of Glasgow.

A further reservoir to the north of Lough Vartry was constructed between 1908 and 1925 to the designs of John George O'Sullivan. The impounding earth dam of the upper reservoir is somewhat longer and lower than the earlier dam. The reservoir covers an area of about 310 acres and has a capacity of around 12 million gallons. The main features are the two large section tunnels under the embankment (29 ft wide by 26 ft 3 in. and 23 ft 3 in. high respectively), and the bell-mouthed overflow shaft, 39 ft deep with a circular ashlar granite weir 72 ft. in diameter. The larger tunnel connects the upper and lower reservoirs, whilst the other accommodates the overflow. The reservoirs are linked by a short steep-sided canal.

The contractors were Kinlen, McCabe and McNally (1908) and H. S. Martin (Dublin) (1919).

NEVILLE P. On the water supply of the city of Dublin. *Min. Proc. Instn Civ. Engrs*, 1874, **38**, 1–149.

O'SULLIVAN J. G. Description of the works in connection with the proposed additional storage reservoir at Roundwood, County Wicklow for the Corporation of Dublin. *Trans. Instn Civ. Engrs Ir.*, 1908, **34**, 94–120.

30. Dublin Water Supply (Blessington)

The growth of the Dublin suburbs in the 1930s resulted in increased water demand and the River Liffey was identified as the source for the additional supply.

When it was learnt that the Electricity Supply Board (ESB) were planning a hydroelectric scheme on the river, Dublin Corporation agreed to become partners with the ESB in the construction of a dam at Pollaphuca in County Wicklow.

This arrangement was legalized in the Liffey Reservoir Act 1936, which initially gave the Corporation the right in perpetuity to abstract around 20 million gallons per day from the reservoir near Blessington. Work on the scheme was carried out (with some difficulties occasioned by war conditions) between 1938 and 1943.

The chief power plant and dam are at Pollaphuca,

while a subsidiary power plant and dam were built 1½ miles downstream at Golden Falls. Together, these power plants have an installed capacity of 34 MW. Three new reinforced concrete bridges of portal frame design, with free spans suspended on cantilevers, were erected to carry roads diverted on account of the creation of the reservoir. The Pollaphuca Dam is gravity-type mass concrete, with a maximum height of 104 ft and a total length of 225 ft. There is a 1230 ft long pressure gallery of 16 ft internal diameter, a 900 ft length being tunnelled in solid rock.

Water is drawn from the reservoir through high-level and low-level tunnels, 390 ft upstream of the dam, which lead to treatment works below the dam at Ballymore-Eustace. From there the treated water enters a 12.3 mile long cast *in situ* ovoid-shaped concrete aqueduct. This terminates near Rathcoole in south County Dublin, from where steel pipelines deliver the water to service reservoirs at Saggart, Belgard and Cookstown on the western fringes of the urban area. The cast *in situ* aqueduct was constructed by the Cementation Company Ltd of Doncaster to designs prepared by the Corporation engineers, under the City Engineer, Norman Chance. The maximum internal dimensions of the aqueduct are 4 ft 6½ in. high by 3 ft 9½ in. wide. A second main was laid in 1953 and a new 25 mile long aqueduct completed to Fairview in the north of the city in 1994.

O'DONNELL K. C. Dublin's water supply. *Trans. Instn Engrs Ir.*, 1987–88, **112**, 105–123.

YOUNG R. M. The Liffey aqueduct. *Trans. Instn Civ. Engrs Ir.*, 1942, **68**, 225–253.

HARTY V. D. The River Liffey hydroelectric scheme. *Conc. & Constr. Eng.*, 1939, **65**, Nov., 581–588.

31. Turlough Hill Pumped Storage Station

Ireland's only pumped storage electricity generating station is situated at Turlough Hill near Glendalough in County Wicklow. It is not a primary producer of electricity, but uses the excess capacity which is available in the national system during periods of low demand – mainly

HEW 3226
T 090 990

ELECTRICITY SUPPLY BOARD

Turlough Hill pumped storage scheme

at night – to pump water from a lower reservoir to an upper reservoir. It then uses this same water to produce electricity during periods of high demand. It can also be used to provide electricity on demand at times of sudden loss of system generating capacity.

The scheme was designed by the Civil Works Department of the Electricity Supply Board. The cavern and tunnels were constructed between 1969 and 1974 by a consortium of German companies led by Alfred Kunz & Co., and in association with the Irish Engineering and Harbour Construction Co. The station was commissioned in 1974.

The upper artificial reservoir, which holds 2.3 million cu. m of water, was formed on the top of the mountain by removing many tonnes of peat overburden. Rock fill was then excavated and used to form a 25 m high embankment, and the inside was lined with asphalt and sealed to make it watertight. The lower reservoir was a natural lake, Lough Nahanagan, the bed of which was lowered by about 15 m. The mean gross head available at the site is 287 m.

The cavern inside the mountain is 82 m long by 23 m wide by 32 m high and entailed the removal of 47 000 cu. m of granite rock. The cavern houses four 73 MW reversible pump turbines and the associated generating units. A single steel-lined pressure shaft connects the turbines with the upper reservoir and has an internal diameter of 4.8 m and a length of 584 m at a slope of 28°.

The tail-race tunnel has a diameter of 7.2 m and is 106 m long, whilst the main access tunnel to the cavern is 565 m long and has a curved segmental section of maximum height 5.5 m above a flat floor 5.7 m wide.

CAROLAN E. J. Concrete structures in lower reservoir, Turlough Hill. *Trans. Instn Engrs Ir.*, 1976–77, **101**, 1–8.

FIVES M. O. Progress at Turlough Hill. *Irish Engineers Journal*, 1972, **25**, no. 6, 15–23.

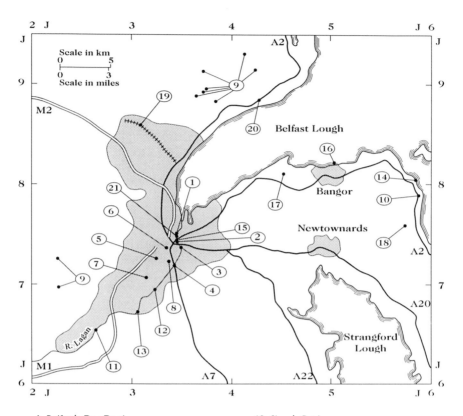

1. Belfast's Dry Docks
2. Queens Bridge, Belfast
3. Albert Bridge, Belfast
4. Kings Bridge
5. Tates Avenue Bridge
6. Boyne Bridge, Belfast
7. King's Hall, Belfast
8. Botanic Gardens Palm Houses
9. Belfast's Reservoirs
10. Hunt's Park and Downpatrick Water Towers
11. Lagan Navigation

12. Shaw's Bridge
13. Drum Bridge
14. Donaghadee Harbour and Lighthouse
15. Lagan Weir
16. Bangor Breakwater
17. Helen's Bay Station and Bridge
18. Ballycopeland Windmill
19. Greenisland Loop Line
20. The 'Horseshoe' Bridge, Carrickfergus
21. Cross Harbour Bridge, Belfast

4. Belfast City and District

In historical terms, Belfast is a very young city, only achieving that distinction in 1888. Belfast lies at the mouth of the River Lagan, on an area which is composed primarily of low quality silty-mud (known as sleech) in places up to 50 ft deep. Most large structures rest on piles, the oldest being of timber. For many years, access to the sea was poor and the primary port was located further down Belfast Lough at Carrickfergus. However, a ridge of firmer ground allowed the fording of the river near a point subsequently bridged by the Long Bridge (now the Queen's Bridge). The origins of the town were laid when a castle was built to protect this crossing, the first charter being granted in 1613.

The growth of Belfast was, however, slow. The site was not ideal for the growth of a large town, with high ground close by on two sides. Development was eventually to follow well defined lines, along the shore of Belfast Lough, up the valley of the River Lagan, and through the gaps in the hills at Dundonald and Glengormley. It was the opening up of access to the sea, however, which was eventually to lead to a massive increase in population. A population of some 71 000 in 1841 had grown to 350 000 by 1900, at a time when houses were being built at a rate of 4000 per year. Large extensions to the city boundary were made in 1896 and again in 1918.

The problem of lack of adequate access to Belfast Lough hampered the development of the port. Both Thomas Telford and John Rennie were consulted by the Ballast Board (set up in 1786), but it was not until 1839 that an improvement scheme was put in hand. This involved the making of a new cut through the mud flats. The work was undertaken by William Dargan and the dredged material was used to make an island beside the cut. Dargan's Island (renamed Queen's Island in 1849) proved to be an ideal location for the development of the shipyards upon which the prosperity of Belfast was to be for so long dependent. Shipbuilding still continues, but the emphasis is now on the construction of large, relatively simple, vessels.

The Lagan Navigation, and the subsequent railway, followed the river valley inland. This led to the development of industry in other parts of Ulster, with Belfast becoming the primary centre for the exportation of manufactured goods. Linen manufacture became particularly important. The failure of the Coalisland mines was to lead to the importation of most fuel supplies through Belfast and a number of smaller ports.

Given the history of Belfast, it is not surprising that many items of its civil engineering heritage are of relatively recent date. There are, however, some excellent examples still surviving of Victorian and Edwardian engineering. The city's water supply, for example, had from 1809 been from a small reservoir (Lysters Dam), the water being fed by an open ditch or leat to Donegal Pass. Head and quantity were inadequate and development of areas in the surrounding foothills necessitated the construction of upland reservoirs, initially in the hills of County Antrim, but by 1900 in the Mourne Mountains. This latter development required the construction of a gravity aqueduct some 40 miles in length.

The rapid growth of Belfast resulted in 1835 in exports through the port exceeding those of Dublin. By 1900, Belfast was the third most important port in these islands.

The city has proved to be an irresistible draw for the rural population. Despite attempts to develop new towns, the greater Belfast area has continued to grow, and is still growing, such that almost two-thirds of the population of Northern Ireland lives within the region.

1. Belfast's Dry Docks

The original port of entry to County Antrim was at Carrickfergus. Although a quay enlargement was commenced at Belfast in 1676, development was hampered by a lack of depth, a problem which remained unresolved until William Dargan cut the first access channel in 1839–41. The shipbuilding industry began on a small scale in 1791, but was able to progress much more rapidly when proper access was provided and following the construction of a series of dry docks (see Table 1). Initially these were on the County Antrim bank of the River Lagan, but in 1853 shipbuilding commenced on Queen's Island (formed on the County Down shore from the spoil dredged by Dargan). This became the nucleus of the extensive shipyards of Harland and Wolff.

William Dargan
(1799–1867), by
Stephen Smith
the Elder

NATIONAL GALLERY OF IRELAND

Table I: Early dry docks at Belfast

Dock	HEW number	Grid ref	Year opened	Length (ft)	Width min/max (ft)	Sill depth below harbour datum
Clarendon 1	1598	J 344 751	1800	235	28	at datum
Clarendon 2	1599	J 344 751	1826	285	34	at datum
Hamilton	1807	J 351 752	1867	450	60/64	5 ft 7 in
Alexandra	1816	J 355 759	1889	788	50/80	15 ft
Thompson	1819	J 357 762	1911	850	96	24 ft 6 in.

HEW 1853
J 362 769

Of the works listed in the table, both of the Clarendon Docks are now preserved in a dry state, but the Alexandra Dock is derelict and likely to be infilled. It will be seen that each dock constructed was progressively larger. When built, the Thompson Dock was the largest in the world and was used by such ocean liners as the Olympic and the Titanic.

By 1962, the Thompson Dock was too small for the large tankers and bulk ore carriers then being built and work commenced on an even larger dry dock. The Belfast Dry Dock was designed by Rendel Palmer and Tritton and built by Charles Brand & Sons between 1965 and 1968. Whilst all of the earlier dry docks are in stepped masonry, the new dock consists of a simple reinforced concrete floor, anchored against uplift, and sheet pile walls. It is some 1150 ft long by 160 ft wide. When built it was second only in size to that at Southampton and one of the five largest in the world.

Table 2 : Designers and contractors of early Belfast dry docks

Dock	Designer	Contractor
Clarendon 1	William Ritchie	William Ritchie
Clarendon 2	Thomas Burnett	H. Mullins & J. MacMahon
Hamilton	W. H. Lizars Assoc.	Thomas Monk
Alexandra	Thos. R. Salmond	McCrea & McFarland
Thompson	Harbour Commission (in-house)	Walter Scott & Middleton

The final major construction work undertaken was the **HEW 2109**
Building Dock. In construction this is very similar to the **J 356 752**
Belfast Dry Dock, but, by cutting off an arm of an existing
channel, excavation was minimized. In this case, the
walls are L-shaped reinforced concrete cantilevers carried on steel piles driven to bedrock. It was completed in
1970 by George Wimpey & Co. to a design prepared by
Babtie Shaw and Morton. The Building Dock marked a
change in shipbuilding practice. Large ship sections
could now be built under cover and then taken to the
dock for assembly. Thus most of the existing timber
sheds, many covered by Belfast truss roofs, were demolished and replaced. Two large Krupp gantry cranes were
installed: dominating the skyline, these are affectionately
known as Goliath and Samson. They have a span of 453 ft
and a headroom of 227 ft and 266 ft respectively. The safe
working load is 840 tons. With these works, ships of
338 000 dead weight tons can now be accommodated at
Belfast docks.

SWEETNAM R. and NIMMONS C. *The port of Belfast 1785–1985*. Belfast
Harbour Commissioners, Belfast, 1985.

ROSS K. *et al*. New dry dock at Belfast. *Proc. Instn Civ. Engrs*, 1972, **51**,
269–294.

2. Queen's Bridge, Belfast

The Long Bridge, built between 1682 and 1688 on the site **HEW 2100**
of a ford over the River Lagan, was, by the end of the **J 344 743**
eighteenth century, in a dangerous state. Accordingly,
the Grand Juries of Antrim and Down obtained a loan of
£28 000 from the Board of Works for a replacement.The
Long Bridge, some 760 ft in length with 1800 ft of approach causeway was not, however, simply replaced. By
infilling on the County Down bank, the spans of the new
bridge, built by Francis Ritchie, were reduced from 21 to
five flat segmental arches, each of about 50 ft span. It was
opened in January, 1843.

Although built to a width of 40 ft, traffic growth resulted in a need for more lanes. In 1886, the City Surveyor,
J. C. Bretland, designed 7 ft wide cantilevered footways
on either side. These are supported by metal brackets

ESLER CRAWFORD, BELFAST

Queen's Bridge, Lagan Weir and Cross Harbour Bridge

carried at the piers by clusters of columns with large capitals.

In 1966 a new parallel bridge, with one major span and half-span side spans, was built in steel. When this, the Queen Elizabeth II Bridge, was opened, the Queen's Bridge (named after Queen Victoria) was made one-way.

McCombs guide to Belfast. S. R. Publications, Belfast, 1861 (reprinted 1970).

3. Albert Bridge, Belfast

HEW 1020
J 349 738

In 1831 a toll bridge (passengers paying one halfpenny to cross) opened at this site. In 1860, the toll was purchased by the local authorities for £4500 and the name changed to the Albert Bridge. The resulting increase in traffic was to cause structural difficulty and in 1886 it was noticed that part was sinking; some 70 ft of the centre collapsed suddenly on 15 September, fortunately at night.

A replacement bridge, designed by J. C. Bretland, was opened on 6 September 1890. This has three flat segmen-

Albert Bridge

tal arches, each with a clear span of 85 ft and each made up of eleven cast-iron ribs. Each arch rib is made up of five segments with intermediate radial stiffeners. The centre arch is to a radius of 99 ft 10in. and the outer arches are radiused to 107 ft 7in. The abutments are in brick with granite ashlar facings (supplied by Newall of Dalbeattie). The contractor was James Henry & Sons of Belfast and the castings were by A. Handyside & Co. of Leeds. The contract price was £36 500.

In 1985 a barge collided with the bridge, causing damage to the outer arch rib on the upstream face. Although of a relatively late date, this bridge across the River Lagan is one of the finest in Ireland built in cast-iron. Rather than replace it, the Roads Service of the Department of the Environment for Northern Ireland let a repair and restoration contract, Dr I. G. Doran & Partners acting as consulting engineers. The elements of the bridge are now picked out with paints of contrasting colours.

BRETLAND J. C. The Albert Bridge, Belfast. *The Engineer*, 1890, **69**, 436.

4. King's Bridge

From 1869 the River Lagan in Belfast was crossed at only four points (one rail bridge and three road bridges). By

HEW 1968
J 340 718

the early 1900s this was proving to be inadequate and in 1909 Belfast City Council advertised for designs for another road crossing. The competition was won by the Trussed Steel Concrete Co., who offered a four-span concrete bridge reinforced on the Kahn principle.

The Kahn system originated in the USA for large factory buildings, but by 1909 it was being promoted for bridge work. The Kahn reinforcing bar was diamond in section and had horizontal extensions which were broken away and bent up (or down) to provide bond and shear protection.

The new bridge, opened about 1912 and named after the new king, is the earliest multi-span Kahn bridge identified in either Britain or Ireland. The distance between the abutments is 195 ft and the width is 30 ft. Minimum headroom is 9 ft 6 in. There are three main beams, 51 in. by 13 in. section for the two 50 ft central spans, reducing to 42 in. by 12 in. for the outer 40 ft spans. It was constructed by W. J. Campbell & Sons.

When constructed, the Belfast Harbour Commissioners (who had jurisdiction to the tidal limit) insisted on a bridge at right angles to the banks to facilitate barge traffic; but as this gives an awkward road layout, plans are in hand to build a replacement bridge alongside.

ANON. Ferroconcrete bridge over River Lagan. *The Engineer*, 1913, **115**, 493–494.

5. Tate's Avenue Bridge

HEW 2143
J 324 726

The bridge at Tate's Avenue is a remarkable mixture of styles and designs. It was built for the Belfast City Council in 1926 to replace a level crossing over what had been the Ulster Railway, which became part of the Great Northern Railway (Ireland) in 1875–76. The bridge is, in essence, three structures. The main, or central, part crosses the railway, but to either side are smaller spans, some infilled with curtain walling to provide storage space, which cross roads running either side of the railway.

The overall width is some 44 ft, and the proximity of existing housing on the Lisburn Road side meant that the two sides were constructed slightly differently. On that side, there is 195 ft of solid brickwork approach ramp,

M. H. GOULD

Tate's Avenue
Bridge

followed by 213 ft with a brickwork core and cantilevered footpaths. These are carried on steel I beams supplied by Glengarnock Colville & Sons Ltd. Spanning onto the I beams is jack-arching, which employed dark blue engineering brick.

The main span over the railway is 78 ft. To the east is a 60 ft span, which is bricked in, and a 39 ft span over the road on this side of the railway. On the west side, there are three spans, the first being 54 ft and infilled, a 39 ft span over the second road, and another infilled span of 42 ft. The crossing ends with a solid brick ramp down to ground level.

The road spans consist of four I beams, the inner two being deeper. The outer beams are carried by massive scrolls, two each end, built into the brick piers. The road surface is carried on steel arched plates, curved at right angles to the centre line of the road. The rail span has two very deep outer I beams. Smaller I beams span to these; and these in turn carry even smaller beams running along the line of the road. For the infilled spans the road is carried on longitudinal beams, but the footpath is again cantilevered out with brick jack-arching.

The main contractor for the bridge was H. and J. Martin. The heavy steel work was undertaken by the Motherwell Bridge and Engineering Co. The iron hand-rail supports, which carry lattice-work handrails, were cast by the Millfield Foundry, Belfast.

6. Boyne Bridge, Belfast

HEW 2092
J 334 737

The old road from Carrickfergus to Lisburn crossed the Blackstaff river at its tidal limit by the Salt Water Bridge, built in 1642 and extensively repaired in 1717. This bridge is named as Brickhill Bridge on Moll's map of 1714.

The terminus for the Ulster Railway was constructed adjacent to the bridge. However, by 1870, the level crossing on the approach was causing considerable delays and the railway company built a three-span road bridge in wrought iron abutting the Salt Water Bridge. In 1934 R. B. Donald, Belfast City Surveyor, designed a wider replacement structure. The resulting bridge, known as the Boyne Bridge (reputedly because, in 1689, the Williamite Army used the Salt Water Bridge on its way to the Battle of the Boyne) is an interesting composite structure.

On the southern approach, parts of the stonework of the old bridge remain, now covered by a layer of reinforced gunite (sprayed concrete). The 60 ft wide by 138 ft long bridge is made up of eight plate girders supporting the carriageway and two parapet plate girders. Supporting uprights consist of six rolled steel joists strapped together. The total weight of steel is 300 tons. The foundation consists partly of precast and partly of cast *in situ* piles and the north abutment was designed as a cellular lightweight concrete construction. The deck is formed by an 8 in. concrete slab. The contractor was H. and J. Martin, and the work was completed on 18 December 1936.

LUCAS F. W. and ALEXANDER R. P. Notes on municipal works in Belfast. *Proc. Instn Mun. & County Engrs*, 1938–39, **65**, 101–110.

7. King's Hall, Belfast

HEW 1632
J 315 707

In March 1933, the journal *The Engineer* announced that Belfast was to have an 'Olympia' style exhibition hall. The design was undertaken by Considère Ltd, London, with

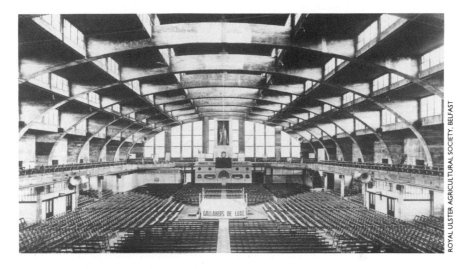

ROYAL ULSTER AGRICULTURAL SOCIETY, BELFAST

King's Hall,
Belfast, as built

A. Leitch & Partners, Architects. The contractor was J. and R. Thompson of Belfast.

The hall is 299 ft long with a clear span of 151 ft. The height is 64 ft. There are eleven three-pin arches in reinforced concrete, each being 6 ft by 2 ft 6 in. section at the base. They are carried up vertically for some 22 ft before turning inwards to follow a compound curve, the section reducing to 5 ft by 2 ft 6 in. The roof consisted of 8 in. thick slabs separated by four vertical windows each side, which gave the building its characteristic profile.

The hall was opened on 29 May 1934 and, with the permission of King George V, was named the King's Hall. It was then the largest ferroconcrete building outside London (said to be second only to the 1929 West Horticultural Hall). In a recess at one end of the hall is a statue of King George VI, purchased from the Ideal Homes Exhibition in London and the first completed of that monarch.

As a result of extensive use of the hall as an ice rink in the 1950s and 1960s, condensation caused considerable damage. In 1983 a new roof structure, designed by Lewis and Baxter, Architects, with Ferguson and McIlveen acting as Consulting Engineers, was installed by H. and J. Martin. The new roof follows the smooth profile of the portals and was installed at a cost of some £750 000; the cost of constructing the original hall had been a mere £61 159!

In the adjacent show ring, part of the original stand, installed in 1896, still exists as the Royal Stand. Beside this is the stand built in Hennebique reinforced concrete in 1920 to a conceptual design by the architect, Thomas W. Henry. The detailed design for this structure, with its tall thin columns, was carried out by Mouchel & Partners.

McCREARY A. *On with the show.* Royal Ulster Agricultural Society, Belfast, 1996.

8. Botanic Gardens Palm Houses

HEW 983
J 337 724

In the late 1830s, Charles Lanyon was engaged to design a glasshouse, using cast-iron sections, for the Belfast Botanical and Horticultural Society. However, Richard Turner of the Hammersmith Works, in Dublin, announced that he was able to manufacture glass in curved strips. Accordingly, a new domed design was prepared.

The two wings, one heated, one not, and each 64 ft by 20 ft by 19 ft 6 in. high, were built in 1839–40 at a cost of £1400. These represent the earliest surviving examples of Turner's work in Ireland. The dome, however, was not built until 1852, when a new design was executed by

Botanic Gardens
Palm Houses

NORTHERN IRELAND TOURIST BOARD

Young's of Edinburgh. This is elliptical in plan, is 37 ft 3 in. high, and cost a further £1000.

The nearby Tropical Ravine House is an unique feature amongst botanic gardens in either Britain or Ireland.

Both buildings underwent major renovation in 1976–83 at a cost of £850 000 and £200 000 respectively. As many as possible of the original wrought-iron and cast-iron elements were reused.

The Palm House at Belfast was probably the first of the large Victorian glass houses to be designed, although both Glasnevin, Dublin, and Kew, London, were completed earlier. Although the Palm House at Belfast is higher throughout, that at Dublin covers a larger area. It is unusual for a country the size of Ireland to have two such well-preserved examples of Victorian palm houses as those at the Botanic Gardens in Belfast and Dublin.

9. Belfast's Reservoirs

The demand for water, as a result of the rapid growth of   **HEW 2137**
Belfast from the mid-nineteenth century, soon outstripped the inadequate water supply then available. To improve the situation, a series of surface impounding reservoirs was constructed in the County Antrim hills. These were built in accordance with proposals that were

Table 3 : Details of Belfast reservoirs

Name of dam	Grid reference	Date of Act	Date opened	Length (ft)	Height (ft)	Reservoir capacity (million gallons)
Woodburn North	J 371 913	1865	1878	2060	41.5	82
Woodburn Middle S.	J 371 891	1865	1865	1780	84	469
Woodburn Upper S.	J 364 888	1875	1868	1615	71.5	367
Woodburn Lower S.	J 378 892	1875	1865	1055	61	107
Dorisland	J 384 881	1875	1878	1070	41.5	66
Lough Mourne	J 413 928	1879	1879	630	14.5	576
Copeland	J 428 915	1879	1879	2030	61.5	133
Stoneyford	J 218 697	1884	1887	3150	38.5	810
Leathemstown	J 215 726	1889	1891	1460	62.5	99

first formulated by J. F. Bateman in 1855. Construction was generally of earthen embankments with clay puddled cores. They are still in use, all but two being classed as 'large'.

All the dams are straight, with the exception of Dorisland, which has a U shape in plan. The shortage of water in Belfast in the late 1880s, because of the massive increase in the industrial base at that time, is shown by the way in which some of these dams were built before the passing of legislative authority. Despite the age of some of Belfast's reservoirs, all still play their part in the water supply to the city.

McCULLOUGH C. W. *Brief history of Belfast water supply.* Belfast, 1923.

10. Hunt's Park and Downpatrick Water Towers

HEW 2113
J 587 789
HEW 2114
J 500 442

At Hunt's Park in Donaghadee is a disused but preserved early concrete water tower. This was built in 1912 for the urban district council. It is, in fact, the larger of two built to this design. The other, which was illustrated in line form in a book published in 1910, is at the Down County Asylum, Downpatrick. Both towers were designed by the Expanded Metal Company and are constructed with expanded metal reinforcement.

At Downpatrick an underground reservoir was built in the same material. Attempts a few years ago to demolish this had to be abandoned as it was found to have considerable strength, even though the expanded metal shows at the surface in several places. Both towers show some interesting detailing which is very reminiscent of earlier towers built in brick.

The contractor for the Hunt's Park tower was J. and R. Thompson of Belfast, whilst the tower at Downpatrick was built by R. D. Pollock. It is believed that the Expanded Metal Company supervised the construction.

11. Lagan Navigation

HEW 2117
J 341 712 to
J 079 627

Work on a scheme to make the River Lagan navigable from Belfast to Lisburn commenced in 1756. The work was directed by Thomas Omer and the navigation was

Hunt's Park
Water Tower

M. H. GOULD

supported by grants from the Irish Parliament. It was opened to Lisburn in 1763 and extended two years later to Sprucefield.

The course of the river was followed throughout. Twelve locks were needed, giving a total rise of 81 ft 9 in.

145

Near the Second Lock, on the Lagan Navigation (painting dated 4 Nov. 1845)

There were no major works on the canal, although at Ballyskeagh (J 289 668) a very fine two arch bridge was built in sandstone. This tall, elegant arched structure is frequently referred to as the High Bridge.

Thomas Omer designed a standard lock-keeper's house, and a number remain in various states of disrepair. Recently two have been refurbished (at Ballyskeagh and Drum).The intention of the original design was for a canal as far as Lough Neagh, and work began again in December 1782, supervised by Richard Owen. Much of the cost was now borne by the Marquis of Donegal. The main structure on this stretch of canal was the aqueduct over the River Lagan at Spencer's Bridge, on which work commenced first. It took three years to complete and cost £3000.

The new cut was formally opened on New Year's Day, 1794. It started at Sprucefield with a flight of locks (Union Locks) giving a rise of 26 ft to the summit. There were ten locks (each 70 ft by 16 ft) taking the canal down to Lough Neagh. Although the most successful of the canals in Ulster, the Lagan Navigation eventually succumbed to other forms of transport. It was abandoned above Lisburn in 1954 and totally in 1958. The canal bed between Sprucefield and Moira was used for the line of Ireland's first motorway (the M1) and the aqueduct was demolished. However, the line of the cut beyond the aqueduct remains and a number of fine stone bridges still exist.

McCUTCHEON W. A. *The canals of the north of Ireland*. David and Charles, Newton Abbot, 1965, 40–61.

12. Shaw's Bridge

This site was for many years the main crossing point on the River Lagan for the road from North Down (and the ports of Donaghadee and Ardglass) to Hillsborough, Dromore and hence to Dublin. Timbers have been excavated, which may be dated to 1617, but any structure which existed here at that time was destroyed during the 1641 rebellion.

HEW 1969
J 325 691

Shawes (with an 'e') Bridge appears on Petty's map of 1653. It is said that during the Cromwellian wars a Captain Shawe had to build a timber bridge to pass his ordnance. In 1698 Thomas Burgh rebuilt it in stone.

This bridge was swept away in 1709. This latter date is sometimes quoted as that of the present bridge, but it had six arches against the extant five. Clearly some or all of the 1709 bridge must have been carried away and again replaced. It is known that the structure was repaired in 1778 and it appears with five arches in a painting dated 1816. Shaw's Bridge highlights the problem of dating Irish stone bridge construction.

There are five near semicircular spans, one of 23 ft 6 in. and four of approximately 30 ft. The pier thicknesses range from 7 ft to 8 ft with triangular cutwaters up to the level of the arch springing. The width between parapets is 16 ft 3 in.

The bridge is now restricted to pedestrian traffic and has been bypassed by a single span concrete arch bridge, built in 1974 and consisting of twelve prestressed portal frames cross-tied below the water line.

M.H. GOULD

Drum Bridge

HEW 2118
J 307 672

13. Drum Bridge

On the outskirts of Belfast, at Drum, is a fine example of a causeway bridge. This was a technique used by the early road builders to cross the flood plain of a river. To reduce the number of arches, a raised causeway, with stone paved sides, was constructed, often, as at Drum, on one side only. The river was then crossed on a conventional, usually multi-span, bridge. Flood relief arches were often provided in the causeway. Dating of such bridges is generally not possible, but Drum was recorded as a bridged crossing point in 1650.

At Drum, the whole causeway is approximately 510 ft long. The three main arches are semicircular and have spans of 19 ft, 22 ft and 19 ft respectively. The piers are about 6 ft thick and have triangular cutwaters to the springing level. To the south are two isolated flood arches with spans of 12 ft and 10 ft 6 in., both now being almost infilled. On the north side of the river are a further four flood arches with spans of between 13 ft and 14 ft. The width between the parapets is 31 ft, but this is the result of at least one widening of the west side around 1802, when John Lamb added 5 ft to the width of the bridge.

The causeway at Drum was later pierced by the Lagan Navigation, but the stone arch has now been badly altered. One of Thomas Omer's lock-keepers' houses may be seen on the west side close by the bridge. Drum was an important bridge on the old road from Belfast to Lisburn (and hence to Dublin). The edges of the road on the Belfast side of the bridge are defined by fine examples of stone walls of the type built around many Irish country estates in the nineteenth century. One wall carries the date, 1841.

Ordnance Survey Memoir, County Antrim, Parish of Drumbeg, 1838.

14. Donaghadee Harbour and Lighthouse

Viscount Montgomery, the local landowner, built a harbour at Donaghadee in 1626. The Post Office decided that it would form the port of call for the mail packets using a new route from Portpatrick (previous sailings had apparently been from Whitehaven). Rebuilding was undertaken between 1775 and 1785, but did not prove wholly

HEW 2110
J 591 801

Donaghadee
Harbour and
Lighthouse

M. H. GOULD

satisfactory. Consequently, a new harbour, with piers to the north and south with a lighthouse at its entrance, was designed by John Rennie Senior; construction was supervised between 1821 and 1836 by his son John.

The south pier is 900 ft long and has four arms or kants. The north pier is 820 ft long and separated from the shore by a shallow channel. The entrance is 150 ft wide between the rounded pier heads. The outer faces of the piers are protected from the action of the sea by rubble pitching at a slope of 1 in 5. The stones used were composed of local greywacke, some of which weighed as much as 10 tons. The inside faces of the piers and the pier heads are curved in a vertical plane and were constructed of ashlar limestone quarried in Anglesey.

Although a steam packet called at the harbour from 1825, it was found that the route was subject to frequent adverse weather conditions and the Post Office service was withdrawn in 1849. In 1856 approval was given for restoring the Donaghadee–Portpatrick route on condition that adequate rail connections were made. The roundabout route to Donaghadee was opened in 1861. However, there was much delay in the completion of the harbour and the rail connection at Portpatrick, which was less than satisfactory.

Steam packets were reintroduced in 1865, but several near mishaps led to a permanent move to Larne in 1867. Here, harbour works had been commenced in about 1845, but no rail connection was provided until 1862. Although considerable amounts of rock were dredged from Donaghadee Harbour between 1858 and 1863 to accommodate the packet steamers, it remains a largely unaltered example of the work of the Rennies. Other examples may be seen at Portrush (1827–36) and Ardglass (1834), but both have been much altered.

GREEN E. R. R. *Industrial archaeology of County Down.* HMSO, Belfast, 1963.

RENNIE SIR J. *The theory, formation and construction of British and foreign harbours.* John Weale, London, 1854, 189–190.

M. H. GOULD

Lagan Weir

15. Lagan Weir

The organic nature of the slob lands in Belfast Lough caused environmental nuisance for many years and, in 1924, Belfast City Council commenced a project to narrow the river. This consisted of some 16 000 ft of pitched embankment topped by roadways, and constructed by direct labour. These are now known as the Annadale, Stranmillis and Ormeau Embankments.When this work was well advanced, W. J. Campbell & Sons were employed to construct a weir and barge lock across the river upstream of its confluence with the River Blackstaff. Known as the McConnell Weir, it incorporated a 40 ft long sluice gate supplied by Ransomes and Rapier of Ipswich. The works were completed in 1937 at a cost of £358 000, including £71 000 for the construction of the weir.

Siltation upstream of McConnell Weir, and continuing problems caused by the slob downstream, led to a reassessment of the situation in the 1980s. Following studies undertaken in 1988, using a large model constructed by

HEW 2116
J 344 744

the Queen's University of Belfast, a new weir was designed by Ferguson and McIlveen and built by Charles Brand Ltd. It is located between the Queen's Bridge and the Cross Harbour Link.

The new Lagan Weir has two functions. First, it holds sufficient water in the river upstream to prevent unsightly mudbanks from being uncovered at low tide. Secondly, by adjustment of the curved gates, it can hold back a flood tide downstream.

Because of the nature of the slob land, extensive foundation work was required for the four river piers. The five 20-ton gates were constructed in the nearby yards of Harland and Wolff. The work was completed in 1994 at a cost of £14 million.

Once the Lagan Weir became operational, the gate was removed and the sill on the McConnell Weir was blown up, the rest of the structure being retained as a feature.

LUCAS F. W. and ALEXANDER R. P. Notes on municipal works in Belfast. *Proc. Instn Mun. & County Engrs*, 1938–39, **65**, 101–110.

MILLINGTON G. S. Development of the River Lagan in Belfast. *Proc. Instn Civ. Engrs*, 1997, **120**, 165–176.

16. Bangor Breakwater

HEW 2095
J 503 822

Like most small harbours in Ulster, that at Bangor has been developed piecemeal over the years. Building of the first pier commenced in 1757. To counteract wave action caused by northerly winds, the North Pier was constructed in 1895, but the problem remained.

In 1977, North Down Borough Council undertook a study of a seafront development scheme, which was to include a marina. Extensive model studies were undertaken by the Queen's University of Belfast, as a result of which it was decided to extend the existing North Pier by some 150 m and to construct a new pier to the west, some 250 m long, to form an enclosed harbour.

Design was undertaken by Kirk, McClure, Morton of Belfast. Conventional rubble-mound breakwaters were considered, but it was realized that there would be a problem with supplying the number of 8–12 tonne boulders needed for the outer faces. Instead, precast concrete 1.2 m cube Shephard Hill Energy Dissipator (SHED)

units, rising from a cast *in situ* concrete toe beam, were used. This type of man-made anti-wave unit had only been used previously on a small scheme in the Channel Islands.

Bangor Breakwater

The contractor, Charles Brand & Co., commenced work on the project in 1979 and continued for the next twelve years. The marina was finally completed in 1994 and the aesthetic appearance of the SHED units has been generally accepted.

KIRK, MCCLURE AND MORTON. *Bangor harbour feasibility study*. Department of Commerce for Northern Ireland, 1977.

HYDRAULICS RESEARCH STATION. *The SHED breakwater armour unit*. HRS report EX 1124, Wallingford, Oxon., 1983.

17. Helen's Bay Station and Bridge

The station at Helen's Bay is something of an architectural *tour de force*, being a miniature example of the Scottish Baronial style. The Marquis of Dufferin and Ava had his estate at Clandeboye, County Down. From the house a private road for carriages was constructed to Grey's Point, some three miles away. This crossed over or under the public roads encountered along the way.

When the railway to Bangor was constructed, it had to

HEW 2112
J 458 821

M.H. GOULD

Helen's Bay
Station and Bridge

cross this road and a very ornamental bridge (rarely noticed by the public) and station were built. Two projections from each face of the bridge pass through the platform walls. This was all designed by Benjamin Ferrey in a style considered 'suitable' and was opened in 1864. The building is now a restaurant.

A short distance beyond Helen's Bay the railway crosses Crawfordsburn Glen by a five arch viaduct (J 466 817), which was built at the same time. It is faced with sandstone, but has brick vaulting.

McCUTCHEON W. A. *The industrial archaeology of Northern Ireland*. HMSO, Belfast, 1980.

18. Ballycopeland Windmill

HEW 2086
J 580 761

The mill at Ballycopeland, County Down, is the only complete windmill in Ulster. This tower mill is believed to date from about 1784 and in 1937 it passed intact into state ownership.

Repairs to combat dry rot and woodworm in the floor and machinery have been necessary from time to time, but it is still remarkably complete. It has a boat-shaped revolving cap carrying the four sail arms and an eight-bladed fantail designed to turn the sails into wind. The

Ballycopeland
Windmill

JACKSON MCCORMICK

motion is carried to a vertical shaft by a bevel gear at the top. The tower, built of local Silurian gritstone, is some 33 ft high and tapers from 22 ft at the base to just under 14 ft at the top. Also on the site is the kiln and the former miller's house. The mill originally produced wheaten, oaten and maize meals using three sets of stones.

RHODES P. S. *A guide to Ballycopeland windmill*. HMSO, Belfast, 1962.

19. Greenisland Loop Line

HEW 2111
J 353 833 to
J 315 854

The Greenisland loop line contains the greatest concentration in Ulster of concrete structures built specifically for the railways. The original railway north of Belfast served the county town of Carrickfergus. From near Greenisland station, the railway to Ballymena branched, going west and then north. Trains from Belfast to the west had to stop at Greenisland, where they either reversed, or the engine was run round the train, causing considerable delays. With the growth of Belfast, and consequential rail traffic to the west, this arrangement resulted in considerable operating expenditure.

The Greenisland
Loop Line under
construction

Between 1931 and 1933 the Northern Counties Committee of the London Midland and Scottish Railway con-

structed a 2¾ mile cut-off line. This was designed to allow trains to the west to proceed unimpeded by services to Larne. The whole scheme was undertaken by direct labour as an unemployment relief scheme.

The works include four under- and four over-bridges, all different, and two large viaducts, all constructed in concrete. The largest structure, which was also claimed to be the largest reinforced concrete railway viaduct in either Britain or Ireland when built, is the main line viaduct. This has three main arches (83 ft clear spans) and nine approach arches (each 35 ft span). At the Belfast end is an additional rectangular opening, which passes the down Larne line underneath as a burrowing junction, being the only one in Ireland. The down line then crosses the down-shore viaduct (one main arch and six approach arches) These two structures contained 17 000 cu yd of concrete.

To the south, the railway tracks cross a road by three parallel square profile bridges, each carrying a different year date. Continuing north is the beam-and-slab Jordanstown Road Bridge, with an unusual three arch profile surmounted by a solid vertical-slab parapet. Monkstown Bridge (recently widened) and its neighbour have rigid frames, and a 58 ft clear span. Auld's Bridge represents an imaginative use of concrete with the railway crossing over a solid barrel arch, 46 ft span. The last bridge at Mossley (also recently widened) has an open spandrel construction with three precast arch ribs. Two rows of precast columns support the deck off the arch. The different designs may be regarded as something of an experiment and these works therefore represent an important stage in the development of the use of reinforced concrete for railway work.

ANON. Railway construction in Northern Ireland. *Civil Engineering*, 1934, **29**, 248–256.

20. The 'Horseshoe' Bridge, Carrickfergus

The concrete bridge at Barn halt on the railway to Larne is locally referred to as the Horseshoe Bridge because of its shape. It was designed by the Belfast office of the

HEW 2097
J 419 879

London Midland and Scottish Railway (Northern Counties Committee) to replace a level crossing, the necessary ramps being constructed on either side of the line and giving it its distinctive shape. It was built in 1928 as part of the doubling of the line between Carrickfergus and Whitehead.

This was the first time in either Britain or Ireland that a bridge was constructed on the 'flat slab' principle. In this form of construction, the 'beams' which span in both directions between the supporting columns are contained within the depth of the slab by changing the density of reinforcement. This technique was first used in the United Kingdom in 1919 for building works, and is still frequently used, for example, in reservoir roofs, but its use in bridge construction was quite unusual. The deck is 10 in. thick and is supported on 39 18 in. diameter columns spaced on a 13 ft 6 in. grid. Each column is on either a 5 ft (outside) or 6 ft square concrete pad. There is a 20 ft carriageway and two 5 ft footpaths.

McILMOYLE R. L. Reinforced concrete railway bridges. *Conc. & Constr. Eng.*, 1930, **25**, 37–45.

21. Cross Harbour Bridge, Belfast

HEW 2115
J 344 745

The cross harbour bridge represents the largest single transportation project undertaken in Ulster to date. There are actually two bridge structures, one connecting the Sydenham bypass to the Westlink (and hence to the M1 and M2 motorways), the other joining the previously isolated railway to Larne to the rest of the system at Central station.

The main road structure, the design for which was approved by the Royal Fine Art Commission, crosses the River Lagan. Known as the Lagan Bridge, it has three spans, an 83 m central span flanked by 55 m side spans. The total road viaduct is 800 m long. The rail structure, Dargan Bridge, is 1400 m long, making it the longest rail bridge structure in Ireland, being over twice the length of the Barrow Bridge near Waterford.

The work was undertaken as a design-and-build contract which was awarded in 1991 to a joint venture formed by Farrans Construction Ltd, Belfast and John

M. H. GOULD

Graham (Dromore) Ltd. The main design was under-taken by Acer Consultants Ltd. Work was completed in August 1994 at a cost of £32 million.

Cross Harbour Bridge, Belfast

The viaducts are supported on 105 land piers, all piled. The deck consists of 1058 matched precast segments, weighing between 50 and 90 tonnes each, built off the piers as balanced cantilevers. This was the first time that matched casting (each unit being cast against its neigh-bour in the casting yard) was used in Ireland. The units were post-tensioned after placement.

DONALD P. T. *et al.* The design and construction of the cross-harbour road–rail links, Belfast. *Proc. Instn Civ. Engrs*, 1996, **117**, 28–40.

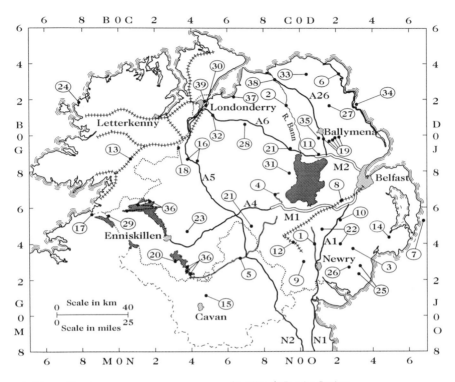

1. Newry Canal
2. Lower Bann Drainage and Navigation
3. Lough Island Reavy
4. Ducart's Canal
5. The Ulster Canal
6. Antrim Coast Road
7. South Rock (Kilwarlin) Lighthouse
8. The Ulster Railway
9. Craigmore Viaduct
10. Dromore Viaduct
11. Randalstown Bridges
12. Newry and Armagh Railway
13. Narrow Gauge Railways of Donegal
14. Quoile Bridge, Downpatrick
15. Ballyhaise Bridge
16. Newtownstewart and Ardstraw Bridges
17. Lennox's Bridge
18. Clady Bridge
19. Mid-Antrim Bridges
20. Inishmore Viaduct

21. Ulster's Dredge Bridges
22. White and Seafin Bridges
23. Drumlone Bridge
24. Cruit Island Footbridge
25. Silent Valley and Ben Crom Reservoirs
26. Spelga Dam Siphons
27. Dungonnell Dam
28. Altnaheglish Dam Stabilisation
29. Erne Drainage and Hydroelectric Development
30. Craigavon Bridge, Londonderry
31. Mullan Cot Footbridge
32. Burntollet New Bridge
33. The Dry Arch
34. Carnlough Harbour
35. Gallaher's Factory Roof, Lisnafillan
36. Concrete Bridges of Fermanagh
37. Number 6 Hangar, Ballykelly
38. Coleraine Bascule Bridge
39. Foyle Bridge

5. Ulster (except Belfast City and District)

Ulster, historically the northernmost province of Ireland, comprises the nine counties of Antrim, Armagh, Cavan, Donegal, Down, Fermanagh, Monaghan, Tyrone and Londonderry.

Ulster contains a wide range of land types, with excellent farming land in the east giving way gradually to poor peripheral land in the west. Lough Neagh is the largest inland lake in either Britain or Ireland.

Deposits of minerals are generally lacking, with only small deposits of coal at Coalisland and Ballycastle, some low-grade iron and bauxite ores on the Antrim Plateau, and gypsum underlying Belfast (from where it cannot be mined). The three most successful mineral finds have been salt at Carrickfergus, the various limestone deposits (used as fertilizer and as a flux in steel making) and gypsum at Kingscourt.

Ulster has always had a rural based economy and close trading links with Scotland. The presence of abundant water power was to lead to the development of spinning and weaving. It was, however, the lack of adequate fuel supplies which resulted in the construction of the first modern summit canal in either Britain or Ireland, linking Newry with Lough Neagh. The scarcity of fuel was also to force many of the larger textile plants to the coastal towns where coal imports were cheaper.

By 1900, Ulster had a reasonable rail and canal network. However, the many conflicts have, by and large, removed any traces of medieval bridge works, with few extant bridges dating from before the mid-eighteenth century. The anti-rail policy of the Northern Ireland government in the 1950s resulted in the closure of much of the network. The replacement road infrastructure, based on the Dungannon motorway (the M1),was built in stages at that time. Rather than opening up the western part of Northern Ireland to development, as had been hoped, further depopulation of this area resulted.

The three Ulster counties (Cavan, Monaghan and Donegal) that remained within the Republic of Ireland have essentially agriculture based economies, although light industries have been established in recent years in many areas. Donegal is the more isolated, but has a well-developed fishing industry and attracts many tourists.

A policy of attracting inward industrial investment, pursued since World War II, has led to the construction of many factories, one of the most interesting designs being at Gallahers at Lisnafillan. In addition to the many road works, water supplies have also received considerable investment during this period.

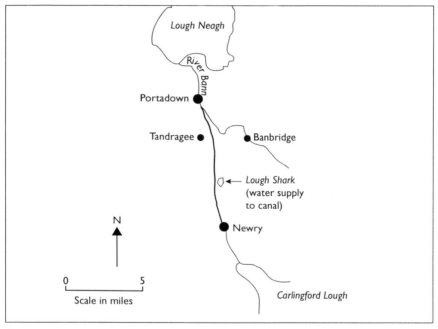

Figure 4. Newry Canal

1. Newry Canal

A canal to link Newry with Lough Neagh was first proposed in 1703, but no action was taken until the Irish
Parliament offered £1000 in 1717 for the first 500 tons of
native coal delivered in Dublin. Accordingly, Edward
Lovett Pearce (Surveyor General) arranged for work to
begin in 1731. Until 1736, the work was in the hands of
Richard Castle (later Cassels). At this point Thomas
Steers (then working on Liverpool docks) took over and
was helped by A. Gilbert. The canal was completed in
1741. This completion date makes the Newry Canal the
first modern summit canal to be constructed in either
Britain or Ireland, the Sankey Brook Canal not being
authorized until 1755.

HEW 1870
J 087 250 to
J 022 526

The canal was 18½ miles long, rose to 78 ft above sea
level, was 45 ft wide and 5 to 6 ft deep. From Newry, there
were nine ascending locks and five descending locks (five
possibly by Steers, the rest by Gilbert). The lock chambers
were 44 ft by 15½ ft and between 12 and 13½ ft deep.

Problems arose as early as 1750, when an attempt was
made to improve the water supply to the canal. It may be
that the Tandragee feeder (which crosses a small stream
by the extant ten arch aqueduct at J 053 446) belongs to
this period.

Accommodation
bridge on the
Newry Canal

M. H GOULD

In about 1800 John Brownrigg reported that the canal was in a ruinous state, including the lock chambers, which had been built of brick faced with Benburb stone and with 2 in. deal floors. Four locks and three bridges were rebuilt, the summit level widened and access to the Newry Ship Canal of 1769 improved. Work was completed in 1811. One bridge, believed to be by Brownrigg, survives at J 064 333, while the bridge at Scarva (J 063 436) is believed to be one of the original structures, although it has been widened on one side.

The canal now entered a period of prosperity, with tonnage doubling in the 1830s to 100 000 tons per annum. In the 1850s it carried 120 000 tons per annum and yielded £4000. Railway competition, however, began to reduce this and by 1918 tolls were no longer covering expenditure. By 1939 the canal was derelict, and it was abandoned in 1956 as a navigation. The course of the canal still exists as a drainage watercourse and it is hoped that it may prove possible to have the navigation reopened.

McCutcheon W.A. *The canals of the north of Ireland.* David and Charles, Newton Abbot, 1965, 17–39.

2. Lower Bann Drainage and Navigation

HEW 2122
H 986 903 to
C 856 303

Lough Neagh is the largest lake in either Britain or Ireland. Several rivers and streams flow into the lough, but only the Lower Bann discharges from it. The total catchment is 2216 sq. miles.

For over two centuries, the flooding caused by Lough Neagh led to repeated calls for remedial action. The first major flood relief works were undertaken in 1840 under the direction of John McMahon, Chief Engineer of the Board of Works, and cost over £250 000. Half of the cost was for the relief of flooding, the other half for making the river navigable from Toome to Coleraine. There were six steps in water level (at Toome, Portna (two steps), Movanagher, Carnroe and the Cutts at Coleraine). Lock chambers were 120 ft by 20 ft. These works improved the drainage considerably. By 1859, however, it was being alleged that the expected relief from flooding was not occurring. No further works were undertaken and it was not until 1925 that Major Shepherd, Chief Engineer of the Ministry of Finance for Northern Ireland, arranged for new surveys to be carried out.

Shepherd found that McMahon's scheme had not been completed. The proposed minimum cross-section of 2400 sq. ft was found in considerable portions to be of less than 900 sq. ft. Shepherd concluded that, as McMahon had died before the completion of the scheme, his successors, faced with increasing costs, had reduced the extent of the scheme.

McMahon had calculated that the weir at Toome would discharge 400 000 cu. ft/min. when Lough Neagh rose to 53 ft ordnance datum (OD). However, with the restricted channel, the Toome weir was drowned at 52.25 ft OD.

In 1930, an order was placed with Ransomes and Rapier of Ipswich for new sluice gates (five at Toome Weir, four with fish passes at Portna and four at the Cutts, all 60 ft wide by 9 ft 6 in. high). The work was undertaken by James Dredging Towage and Transport of Southampton, who sublet the rock excavation and weir and sluice work to Walter Scott & Middleton of London.

Since the completion of Shepherd's scheme, which established a water level of 54 ft OD, there have been two further lowerings of the Lough, each time achieved by adjustment of the sluice gates. The level is now 50 ft OD.

SHEPHERD MAJOR W.E. *Chairman's address to Belfast and District Association of the Institution of Civil Engineers*, 29 Jan. 1934.

Public works and progress. *Civil Engineering. 1932,* **27,** March, 39–42.

3. Lough Island Reavy

The Bann Reservoir Company was formed in 1835 by a group of mill owners anxious to protect their source of motive power. William Fairbairn (who employed J. F. Bateman to make the surveys) was asked to report. He proposed the construction of three reservoirs to be used to control the flow in the river. His proposal was adopted under the terms of the Bann Reservoir Act of 1836, but only two reservoirs were constructed, at Lough Island Reavy (completed 1839) and at Corbet's Lough (completed 1847).

HEW 2120
J 295 340

At Lough Island Reavy, the original lake of 92 acres was extended to 253 acres by the construction of two small banks (516 ft and 648 ft long) and two major banks (1605 ft and 2126 ft long) aimed at raising the water level by 35 ft.

The bank construction was straightforward, being earth dams with puddle cores (12 ft thick at the maximum height of 40 ft tapering to 8 ft). An unusual feature was the inclusion of 3 ft of dry peat either side of the puddle. The water face also had 3 ft of peat, overlain by 18 in. of gravel and 2 ft of pitching. The intention was that the water would swell the peat and so seal any leaks. The banks sloped at 1 in 2.5 (1 in 3 below 20 ft) on the water face and 1 in 2 (1 in 2.5 below 20 ft) on the grassed side. The works were undertaken by William Dargan at a cost of £12 300.

Through the dam was a culvert with a solid central masonry wall and carrying three 18 in. diameter pipes. Later practice would have been different. This culvert was to be puddled against the bank, but over-excavation by the contractor, back-filled with rubble, resulted in the puddled culvert joints failing under 12 ft of water. Re-cauking with oakum and cement was undertaken in 1839–41, but the culvert still leaked. In 1867, a length of culvert by the wing wall was exposed, poor material removed, and the length was repuddled. Some defective joints were remade and the leak was thus eventually sealed.

At Corbet's Lough, flow control weirs were installed together with an embankment which raised the water level by 12 ft. These two reservoirs were estimated to have raised the hydropower potential of the Upper Bann river by a factor of five.

Following the drought of 1973, and given that most of the mills remaining along the river had ceased to use water power, a new eduction tower and overflow were built and Lough Island Reavy was connected to the Silent Valley Aqueduct for water supply purposes.

BATEMAN J. F. Bann reservoirs. *Min. Proc. Instn Civ. Engrs*, 1848, **7**, 251–274.

SMITH J. JNR. A long existing leak and its repair. *Trans. Instn Civ. Engrs Ir.*, 1869, **9**, 51–67.

4. Ducart's Canal

The mining of coal at Coalisland in County Tyrone was a long established tradition, although the seams show severe faulting. Transport of the coal was a problem, but was relieved by the opening of the Coalisland Canal in 1733. In 1753, proposals were made for an extension of the canal to Drumglass, and this was completed in 1777.

HEW 2132
H 804 645 to
H 844 665

The method of working of the extension, usually referred to as Ducart's Canal, after its designer Daviso de Arcort (Davis Ducart), was unique in Ireland. Ducart was a Sardinian architect, whose works in Ireland include the Custom House in Limerick. The coal was conveyed in small wheeled tubs, which were run on to lighters on the canal. This consisted of three level sections connected together and to the existing canal by stone arched ramps, known as 'dry hurries', down which the tubs were run to the next level.

Coal output was less than expected and Ducart's Canal soon fell into disuse. Parts of the hurries remain, the

Newmills
Aqueduct,
Ducart's Canal

M. H. GOULD

one most easily seen being at Furlough beside the Dungannon to Newmills road at H 812 665.

The most prominent work still extant is the ashlar stonework aqueduct which carried the canal over the River Torrent at Newmills (H 815 673). This has three near semicircular arches and remains in an unaltered state. It now provides vehicular access to Newmills sewage disposal works.

McCutcheon W. A. *The industrial archaeology of Northern Ireland*. HMSO, Belfast, 1980, 58–65.

5. The Ulster Canal

HEW 1888
H 419 210 to
H 854 561

The Ulster Canal was the last major line of canal to be constructed in Ireland. The plan was for boats to travel from Belfast (via the Lagan Navigation) and from Newry (via the Newry Canal) to Lough Erne and thence by the Ballinamore and Ballyconnell Navigation to the River Shannon. However, by the time it was completed, the Ballinamore and Ballyconnell Navigation had yet to be built. Furthermore, for reasons which have never been established, the Ulster Canal was completed with locks that are considerably narrower than any other in Ireland (approximately 12 ft wide).

The first survey of the line of the navigation was undertaken by John Killaly, who proposed using the standard adopted for the Royal Canal, with locks 76 ft by 14 ft (somewhat smaller than both the Lagan Navigation and Newry Canal). A revised survey in 1825 called for 18 locks. The Irish Loan Commissioners consulted with Thomas Telford, and on his recommendation, the number of locks was increased to 26. Supervision of the work was undertaken by William Cubitt and the contractor was William Dargan. There are 19 locks (each with a rise of 8 ft 6 in.) between Lough Neagh and the summit level, and a further seven locks down to Lough Erne. The last lock, at Woodford, was only 11 ft 8½ in. wide. The work was completed by 1841.

The canal was never a success. Mention has already been made of the narrow locks. A problem also arose with the water supply and the canal was closed between 1865 and 1873, while remedial work was undertaken. In fact, the Ballinamore to Ballyconnell link only existed during this period (from 1860 to 1869), so the concept of a through route never materialized.

In 1888, the more successful Lagan Navigation Company was persuaded to take over the Ulster Canal, but found it a millstone around its neck. Parts of the canal beyond Clones in County Monaghan became disused, and no boats used the lower section after 1929. This led to formal abandonment in 1931.

Most of the locks still remain, but are generally infilled. However, parts of the line are either ploughed out or blocked. Even so, a feasibility study into the possibility of rebuilding the canal has recently commenced.

McCutcheon W. A. *The canals of the north of Ireland*. David and Charles, Newton Abbot, 1965, 98–119.

6. Antrim Coast Road

The Antrim Coast Road was one of the first major lines of road built under the provisions of the Public Works (Ireland) Act of 1831, being some 33 miles long. The object of the road was to open up the coastal towns previously

HEW 1967
D I55 407 to
D 407 035

Black Arch,
Antrim Coast
Road

M. H. GOULD

served by steep mountainous roads, largely impassable in winter, many of which still exist.

The design, prepared by the Board of Works engineer, William Bald, was undertaken in sections, as evidenced by the loans for the work.

Where it runs beside the sea, the road was constructed by blasting rock from the overhanging cliffs onto the foreshore and building on the resulting berm. These sections have constantly given trouble from wave action. In 1967, for example, a cliff fell onto the road near Glenarm, followed a few months later by a larger rock fall. Some 2 miles of road were reconstructed on the seaward side as a result.

The building of the road over Carey (now Cushleake) Mountain was a magnificent feat. Bald undertook the progressive drainage of deep peat to allow work to proceed (on this section is the famous 'vanishing' lake). A similar technique was subsequently used for building early sections of the Midland Great Western Railway.

D 215 322

The most significant structure on the road is the soaring three-span Glendun Viaduct. Although the overall concept of the road was undoubtedly Bald's, the detailed design for Glendun is widely attributed to Charles Lanyon, when he was County Surveyor of Antrim.

Most of the stone bridges still remain, each with the name, date and Ordnance Survey bench mark cut into the parapet. The bridge at Corratavey is some 47 ft high and into its parapet are cut the words 'William Bald—Engineer'.

The section to the north of the Glendun Viaduct was cut into the side of the hill and is, in parts, held up by 15 ft high stone retaining walls, with inward piers, such that the weight of the road holds down the whole wall. Near Larne the road cuts through a short length of cliff at Blackcave Tunnel (D 398 053).

First, second and third reports of the Board of Public Works (Ireland). H. C. Printed Reports, 1832–34.

TURNER R. M. The reconstruction of the Antrim coast road. *Paper to NI Association of Institution of Civil Engineers,* 10 Jan. 1972.

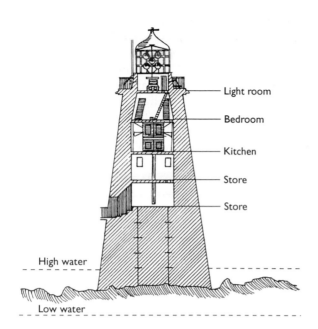

Section through
South Rock
(Kilwarlin)
Lighthouse

Light room

Bedroom

Kitchen

Store

Store

High water

Low water

7. South Rock (Kilwarlin) Lighthouse

This tower lighthouse was described in 1830 as being 'a **HEW 2098**
station of great importance'. South Rock and the nearby **J 677 532**
North Rock lie off the Ards Peninsula and present a
hazard to coastal shipping. Building commenced at
South Rock in 1795 to a design by Thomas Rodgers,
contractor to HM Revenue Commissioners and took a
number of years to complete.It was built of granite
blocks, the end joints of the tiers being joggled into each
other. Eight iron bars, 3 in. square, run within the wall,
being keyed into circular plates every third or fourth
course. The tower is 37 ft diameter at the base, tapering
to 17 ft under the cornice at a height of some 70 ft. The
first 22 ft is of solid masonry.

The design for the lighthouse was considered to give
great strength. It pre-dates Robert Stevenson's taller Bell
Rock (1807–10) and is said to have been the basis of
Stevenson's design.

In order to assist in the transport of Mourne and

Wexford granite to the site, a small harbour was built nearby of local Silurian stone (J 647 533). As this is slaty in nature, the walls were made of small-depth pieces set vertically. This construction remains remarkably intact. Beside the harbour a house was built for the keeper. In 1810, responsibility for Irish lighthouses was vested in a new authority, the Corporation for Preserving and Improving the Port of Dublin (commonly known as the Ballast Board). At this time, a further block of three houses was built to seaward.

South Rock had been used to give a foundation for the lighthouse but it is only about 2 miles off shore. Despite the single revolving white light, wrecks continued to occur. Accordingly, the light had to be abandoned and a lightship stationed a further 2 miles out from the lighthouse.

'Lighthouses', *The Edinburgh Encyclopaedia*, 1830, **13**, 17.

LONG B. *Bright light, white water*. New Island Books, Dublin, 1993, 187–189.

8. The Ulster Railway

HEW 1898
H 873 457 to
J 335 737

The second railway line to be constructed in Ireland was the Ulster Railway, a 36 mile long line from Belfast to Armagh. Built as a single 6 ft 2 in. gauge track using longitudinal sleepers, it opened to Lisburn on 12 August 1839. Lurgan was reached in 1841, and Portadown the following year. The line was doubled and converted to the Irish gauge of 5 ft 3 in. Between 1846 and 1849 it was extended to Armagh.

The construction of the line posed no major problems. Between Belfast and Lisburn a number of fine original bridges built of sandstone blocks still remain. Of note is the bridge over Upper Dunmurry Lane (where the embankment was later pierced by a pedestrian subway), the river crossings at Glenburn and Conway, and the two road crossings at Lambeg.

On the outskirts of Lisburn is another fine structure, which carries the former turnpike road over the railway at Wallace Park, although the original stonework is now masked by later stone facing.

Only one of the original stations remains unaltered,

that at Moira (J 157 618). This is a two-storey construction with accommodation for the stationmaster located below the platform. The station is now in state care.

Near Portadown, the contractor encountered difficulties with wet soil, necessitating special treatment so that it could bear the weight of ballast and track. There were also problems with securing the bridge over the River Bann. It was originally built in timber, but a metal structure was built in 1871, which has since been replaced at least twice.

Thomas Jackson Woodhouse, County Surveyor of Antrim, designed many of the original sandstone bridges. The contractor, William Dargan, was supervised by John Godwin, and both William Bald and George Stephenson also advised the railway company.

McCUTCHEON W. A. *The industrial archaeology of Northern Ireland*. HMSO, Belfast, 1980, 104–107.

CASSERLEY H. C. *Outline of Irish railway history*. David and Charles, Newton Abbot, 1974, 22, 168–169, 181.

9. Craigmore Viaduct

The railway line between Dublin and Belfast was constructed in sections by three different companies over the period 1837–55. An Act passed in 1845 gave powers for the construction of the last section between Drogheda and Portadown by the Dublin and Belfast Junction Railway Company. Through running between Dublin and Belfast finally became possible in 1855 when the Boyne Bridge and Viaduct at Drogheda were completed.

HEW 1973
J 067 284

The section of about 6 miles through the high ground west of Newry involved a considerable amount of rock cutting. It also included a number of masonry skew arch bridges of remarkably thin section, the Egyptian Arch (one of only a few structures in Ireland built in the Egyptian style then in vogue) and the impressive Craigmore Viaduct. The last named was opened to traffic on 13 May 1852.

The Craigmore Viaduct, which is on a slight curve at its northern end, has 18 semicircular arches, each of about 59 ft 6 in. span. The piers are 7 ft 6 in. thick at springing level. The width of the trackway is 28 ft. The height of the

JACKSON MCCORMICK

Craigmore
Viaduct

track above the lowest part of the Camlough river valley
is 137 ft, making it the tallest railway bridge in Ireland. It
was built of local Newry grano-diorite and in 1851–52
cost about £50 000. The viaduct was designed by Sir John
Macneill and built by William Dargan.

An original contract drawing shows a tramway under
one of the central arches of the viaduct, which ran as the
Bessbrook and Newry Electric Tramway between 1885
and 1948. The new station for Newry, which was opened
some years after the closure of the Goraghwood to Newry
line in 1965, is located near the southern end of the
viaduct.

MORTON R. G. *Standard gauge railways in the north of Ireland.* Belfast
Museum and Art Gallery, 1962.

ROBB W. *A history of Northern Ireland railways.* Northern Ireland Railways, Belfast, 1982.

10. Dromore Viaduct

HEW 1899
J 195 533

Dromore Viaduct, a seven-span single-track railway viaduct, was built in 1860–61 for the Banbridge, Lisburn and
Belfast Railway. The viaduct has ashlar stonework, with
piers tapered to a common level and is on a curve over
the River Lagan at Dromore, County Down. The key-
stones of the arches are carried as near semicircles. The

facing stonework is carried up to form low parapets above the track bed.

The total length measured along the inside of the curve is 333 ft. On this face, the spans are 36 ft and the piers 5 ft wide. The width between the parapets is 17 ft 6 in. and across the parapets 22 ft 6 in. The parapets stand 3 ft high to the top of the large flat coping stones.

Since the closure of the line in 1956, the viaduct has become an isolated structure with top access denied, although lying in a riverside park. The new Belfast to Dublin bypass road passes close to the outer face of the viaduct, which has been floodlit by Banbridge District Council. The council has also undertaken waterproofing of this listed structure.

11. Randalstown Bridges

At Randalstown, County Antrim, is a striking railway viaduct in cut stone. This was built as part of the extension of the branch railway line to Randalstown from the Belfast and Ballymena Railway. The extension was to Cookstown and was completed in 1855–56.

HEW 2130
J 084 902

The viaduct has basalt stonework piers and facings to its eight arches, the barrels themselves being completed in brick. The track was single, although the structure appears to be wide enough for a double track, with an overall width of 31 ft 4 in., compared to 22 ft 6 in. at Dromore. A number of structures on this line appear to have been built with doubling in mind—there is an adjacent river bridge with double width piers carried only to water level. The eastern approach crossed the Belfast road by a girder bridge, which has been removed since the line closed in 1959. The first arch crosses a (now disused) mill-race, the next six cross the River Main and the last is over a bankside road. Spans are each approximately 30 ft and the height of the parapet over the river bed is 52 ft. The design was by Charles Lanyon and the contractor was William Dargan.

In the shadow of the viaduct is one of Ulster's fine, almost undatable, multi-arch stone road bridges. This is at an early crossing point on the River Main and the road here was one of the Irish turnpikes. It is stated in the

M.H. GOULD

The Randalstown Bridges

Ordnance Survey Memoir that this bridge is a 'clumsy old structure of some antiquity'. It was twice widened (lastly in 1816), consists of nine semicircular arches of unequal span and is 193 ft long by 29 ft wide.

12. Newry and Armagh Railway

HEW 2121
J 052 349 to
J 038 356

The Newry and Armagh Railway obtained an Act in 1857 to continue its line to Armagh from Goraghwood via Markethill. Traces of the earthworks of this single line railway may still be seen. The railway was never built beyond Armagh and, in 1879, the line was absorbed as a branch of the Great Northern Railway (Ireland).

The 16¾ mile line had sharp curves and steep gradients and involved four different contractors. The contractor in 1861 for the Markethill to Goraghwood section was Robert Samuel North of London. The line was opened on 25 August 1864 to a temporary terminus one mile short of Armagh, the junction being completed the following year. Part of the construction of the line was financed by a grant from the Board of Works as a means of providing unemployment relief.

The line was never a financial success and, by 1933, it was cut back to a goods only branch as far as Markethill. On this section from Goraghwood to Markethill was the longest railway tunnel in Ireland, being the 1759 yd bore at Lissummon. Although the branch was closed in 1955, the tunnel still exists, although now largely derelict. For some years, a new use was found for the western end of the tunnel – the growing of mushrooms.

There has recently been some discussion about the possibility of including the tunnel in a footpath or cycle path, but as yet no detailed survey has been carried out. It is believed, however, that the masonry wall and brick arching to the tunnel are still in good condition.

McCUTCHEON W. A. *The industrial archaeology of Northern Ireland*. HMSO, Belfast, 1980.

ROWLEDGE J. W. P. *A regional history of railways, Volume 16: Ireland*. Atlantic Transport Publishers, Penryn, Cornwall, 1995.

13. Narrow Gauge Railways of Donegal

Unfortunately, little remains of what was once the most extensive narrow gauge railway system in either Britain or Ireland. Only the piers remain of the viaduct at Owencarrow (C 072 266), but some lengths of trackbed may still be found, most noticeably from Stranorlar through the Barnesmore Gap, opened in 1882 (C 088 252). The system developed as two separate entities, cross-connection eventually becoming possible at Londonderry and Letterkenny.

HEW 3249

In the north of the county was the Londonderry and Lough Swilly Railway (L&LSR) and in the south, the Donegal Railway Company (DRC). These ran from Londonderry to a pier on Lough Swilly and, as the Finn Valley Railway, from Strabane to Stranorlar respectively. Both opened short lengths of Irish gauge track in 1863. However, when the extension of the L&LSR to Letterkenny (C 174 118) was left uncompleted because of financial problems, the companies changed to the 3 ft gauge (first used in Ireland in Glenariff, County Antrim in 1876).

Considerable government support was provided for the construction of many of the 3 ft gauge lines. Principal branches were: L&LSR: Letterkenny, 1883; Carndonagh, 1903 (cut back to Buncrana in 1935); and Burtonport, 1903 (B 717 152); CDR: Killybegs, 1893; Glenties, 1895 (G 820 945); Ballyshannon, 1905; and Letterkenny in 1909. A narrow gauge line from Strabane to Londonderry was opened in 1900.

These railway companies undertook pioneering work with regard to the provision of locomotive power on the narrow gauge. In 1931, the companies pioneered the use of diesel (as opposed to petrol) railcars.

Despite these measures, both systems were to suffer from low revenues from the sparsely populated hinterland, a collapse of fishing during World War I, and competition from road transport. The vestiges of the County Donegal railway system were closed down in 1960. Proposals are made from time to time to relay part of the track, especially the steeply graded section to the Barnesmore Gap, but to date, little has been achieved.

PATTERSON E. M. *The County Donegal railways*. David and Charles, Newton Abbot, 1962.

PATTERSON E. M. *The Lough Swilly Railway*. David and Charles, Newton Abbot, 1964.

14. Quoile Bridge, Downpatrick

HEW 2099
J 488 465

The Quoile Bridge stands beside a long established ford at Baghil on the eponymous river. McCutcheon has stated that there has been 'a bridge since about 1640, but the present bridge, which has been very substantially rebuilt, dates from about 1680'. What 'substantially rebuilt' means in this context is unclear. The three-centred arches and triangular cutwaters are very similar to Clara Bridge in County Wicklow, which O'Keeffe and Simington (1991) date to somewhere around 1700.

The Ordnance Survey Memoirs describe Quoile Bridge as being 312 ft long (201 ft over the river), 24 ft wide, and consisting of six arches with scarcely any rise (assumed to mean no hump in the road level). The Memoirs remark that the bridge is 'extremely old'.

This bridge typifies the problems associated with the accurate dating of many of Ireland's fine stone bridges.

Quoile Bridge,
Downpatrick

McCutcheon W. A. *The industrial archaeology of Northern Ireland*. HMSO, Belfast, 1980, Plate 6.

15. Ballyhaise Bridge

Ballyhaise has probably the oldest bridge in County Cavan. The castle at Ballyhaise was built as an extension of the plantation settlement at Cootehill in the early years of the seventeenth century, and local tradition has it that the bridge was built across the Annalee River around the same time. It certainly existed when a new house near the bridge was shown in a painting completed shortly after the house was built in 1705. It would be in the logical position for a road connecting the defensive castle at Ballyhaise with Cootehill and other settlements to the north. The bridge has a stone plaque, but it has proved to be unreadable. In style, however, the bridge does appear to date from the early seventeenth century.

HEW 3247
H 450 112

The bridge is a narrow stone structure with seven segmental arches of 19 ft span and one of 10 ft. It is 16 ft wide overall and has triangular cutwaters carried up to the parapets to form refuges. These, rather unusually, are

179

not all carried up, and may be, in part, a consequence of later alterations; otherwise the bridge appears to be in its original state.

16. Newtownstewart and Ardstraw Bridges

At Newtownstewart in County Tyrone, and at nearby Ardstraw, there are two fine multi-span stone arch bridges built in rubble work and dating from the early eighteenth century. In both cases, the arches are semicircular and triangular cutwaters are carried to the soffit level of the arches.

HEW 2140
H 403 858

The bridge at Newtownstewart, over the River Strule, has six arches. Fine views of the bridge, which is on a minor cross-country route, can be had from the main road from Omagh to Strabane. On Moll's map of 1714 the road is shown as crossing the bridge here before proceeding on the opposite bank to Strabane, then apparently served by ferries from both the south and the west (later sites of bridges).

Newtownstewart Bridge has been largely bypassed by the adjacent Abercorn Bridge, a concrete structure designed by Mouchel & Partners, and opened in 1932.

HEW 2139
H 349 873

At Ardstraw, the four main arches carry what is now another minor road over the River Derg. However, in 1714 the main road to Londonderry crossed here apparently by a ford, before proceeding on to Clady Bridge. The arches are approximately 12 ft span and the piers are 5 ft wide. On one bank there is a wider buttressed pier followed by a further arch providing flood relief. Ardstraw Bridge has been dated to 1727.

17. Lennox's Bridge

HEW 3250
G 816 574

The Drowes river is crossed four times in its course from Lough Melvin to the sea. Beside the sea is Bundrowes Bridge (now bypassed with a modern structure), whilst near the Lough is Mullanaleck Bridge. The third bridge (Lennox's) is approximately one mile downstream, and has the appearance of being the oldest.

Moll's map of 'main routes', published in 1714, shows

Lennox's Bridge

the road from Bundoran to Sligo as crossing Bundrowes, but then taking a more inland route than the present one. This road is also on Taylor and Skinner's 1778 map which picks up the ends of the road from Ballyshannon to Manorhamilton (that crossing Mullanaleck) as an unsurveyed route. Neither shows a road from Bundoran to Manorhamilton.

In appearance Lennox's Bridge is very narrow and humped. It has six arches (with a maximum span of 14 ft 9 in.) with cutwaters forming pedestrian refuges. The profiles of the arches are irregular, with two exhibiting a distinct point, in a manner very reminiscent of the bridge at Killaloe in County Clare, parts of which have been dated to 1690. By contrast, Mullanaleck, which also has pedestrian refuges, is flatter with three more regular arches (max. span 21 ft 4 in.) and probably dates from the early eighteenth century.

18. Clady Bridge

The River Finn (flowing largely in County Donegal) is spanned by a series of old stone bridges downstream of Stranorlar. The most interesting of these, however, is that at Clady (on the border between Counties Tyrone and Donegal).

HEW 2133
H 294 940

M. H. GOULD

Clady Bridge

Clady Bridge has all of the characteristics of medieval construction. There are nine arches of differing size, each with a near semicircular profile. The piers are remarkably wide with triangular cutwaters which are carried up to the parapet level. The bridge is built to a flat profile. The road is very narrow (12 ft), but the refuges formed by the cutwaters are very deep, some 11 ft either side of the roadway.

No construction date has been assigned for this bridge. However, the fort at Culmore used to be an important entry point for the military, and it is known that the road passing down the River Foyle served several inland forts, going to Omagh in the east and Beleek and Ballyshannon in the south. Clady Bridge is shown on Moll's map of 1714 (which shows no bridge downstream at Lifford) and it seems safe to assume that this was where this road crossed the river. The River Derg had to be crossed at Castlederg by the same route and it is on record that a bridge existed there from the first quarter of the seventeenth century, so it seems most likely that Clady Bridge is of a similar date.

19. Mid-Antrim Bridges

The Kells Water flows in a westerly direction across the centre of County Antrim to join the River Main to the south of Ballymena. It is crossed by a series of bridges which epitomize the type and style of such structures throughout Ulster. The number of spans ranges from three to seven, with none exceeding 30 ft. Dating of such bridges is notoriously difficult because of the lack of written records.

The need to cross the wide flood plains adjacent to many rivers caused problems for the early bridge builders. Slaght Bridge (J 090 997), over the River Main near its confluence with the Kells Water, illustrates one solution, with its nine small arches over the flood plain leading to the main four arch structure. Rock Bridge (J 160 983) shows an alternative design, with the main structure approached by a formally raised embankment with paved sides. It has given periodic trouble and is propped up by additional piers of irregular size and spacing. Both appear of some antiquity, but are not shown on pre-Ordnance Survey maps.

Shank's Bridge (J 125 981) is a good example of early nineteenth-century work. Until at least 1778 the road went by Kells (where the old bridge was swept away on

HEW 2138

Battery Bridge

M. H. GOULD

5 November 1834). However, the main road crossed Shank's by the time the first OS map was issued in the 1830s. In the accompanying Memoir, it was remarked that a new turnpike (on the line of the present main road) was in the course of construction. The bridge is described as having seven semicircular arches and being 130 ft by 22 ft. This is either an error of transcription, or the 'over water' length, as the total length is nearer 170 ft.

Moorfields Bridge (J 192 994) was built to connect a feeder road to the line of new road from Ballymena to Larne (for which a loan of £12 200 was given in 1841). Moorfields owes more in its conception to the railway engineer, having almost the proportions of a railway viaduct. Indeed, a 3 ft gauge railway, opened in 1878, ran originally under one approach span, and Moorfields station building still exists in the shadow of the bridge.

Battery Bridge (J 220 990) is one of only a very few with rather flat pointed segmental arches. This is usually an indication of an early bridge. It does not appear on Moll's map of 1714 (which only covered the main routes), but is on Taylor and Skinner's map of 1783. It was described in the Ordnance Survey Memoir as being in 'generally very bad' order, suggesting some antiquity.

20. Inishmore Viaduct

HEW 2125
H 254 342

The road from Carry Bridge crosses to the west shore of Upper Lough Erne on one of only a very few metal road viaducts in Ulster. The road surface is carried on steel arched plates, which span from the longitudinal girders at right angles to the road centre line. It was designed by J. Price and built by direct labour by the promoter (J. V. C. Porter). Construction commenced in 1891, but difficulties were encountered, especially in the construction of the embankments, and the bridge was finished in 1900 by the contractor, Andrew Mair. The final cost of some £5000 was twice the estimate, much of the cost apparently being borne personally by Porter.

The upper cross bracing was strengthened and raised after it was hit by a lorry, giving the viaduct its present top-heavy appearance. The deck was also strengthened by the addition of concrete above the arching.

Nearby, the road to Cleenish Island is carried on a three-span Callender–Hamilton Type B galvanized sectional steel bridge, erected in 1953. This type of structure is usually considered as temporary, but there are no plans to replace it; indeed a concrete deck was added in 1986.

HEW 1972
H 257 385

GOULD M. H. Bridge of the month. *Ulster Architect International*, 1996, **13**, no. 13, April/May, 32 and **16**, no. 6, July/August, 36.

21. Ulster's Dredge Bridges

Ulster originally had several bridges built to the patent of James Dredge of Bath, but these have now mostly disappeared. There was only one which was not a footbridge—that at Ballievy, near Banbridge. This was totally destroyed in 1988 when a large lorry tried to cross it, even though it was signposted as having a 2 ton weight limit. The Dredge patent, whilst very innovative for its time, is, unfortunately, largely unsuitable for modern loadings.

The Dredge system used wrought-iron suspension chains, which increased from a single bar at the centre by one at each link position back to the support. Thus the longer the span, the more elements there were in the cable. This arrangement had the advantages of less weight and a shorter erection time. The suspension

Glenarb Dredge
Bridge

M. H. GOULD

chains pass over the cast-iron support to an anchor behind. The elements are connected together with bolts and shear pins.

HEW 1857
H 758 447

At Glenarb, Caledon is a Dredge patent suspension bridge which was cast by the Armagh Foundry and opened originally in 1844. The main span is 77 ft 9 in. with side spans of 19 ft 9 in. The width is 3 ft 6 in. Glenarb Bridge is unusually narrow for a Dredge bridge and exhibits a more elaborate bracing system than that usually seen. Even so, the bridge is very flexible with a noticeable movement underfoot.

The site was to be affected by a River Blackwater drainage scheme, so in 1984 the bridge was taken down, refurbished, and re-erected in 1990 on new foundations at its present location.

HEW 1856
H 745 432

The estate of Lord Caledon, whose millworkers the bridge originally served, contains a second Dredge bridge. This has a single 73 ft 2 in. span on masonry abutments and a deck width of 10 ft 6 in., and accommodated horses and carts. Also cast by the Armagh Foundry (in 1845), this bridge is not at present open to the public.

HEW 1858
H 927 936

At Moyola Park, Castledawson, there were originally two more Dredge footbridges. One of these, built about 1846, was swept away in 1929. The remaining example has two 66 ft spans either side of a single masonry pier on an island in the river. It is believed that this was the only bridge built by Dredge which had two back-to-back half catenaries. It dates from 1847 and has a width of 5 ft 9 in. It is in poor order, but at the time of writing plans are in hand to try to restore this important grade A listed structure.

McQUILLAN D. Dredge suspension bridges in Northern Ireland: history and heritage. *The Structural Engineer*, 1992, 7 April, **70**, 119–126.

22. White and Seafin Bridges

The Upper Bann river is crossed by two cast-iron bridges with brick jack-arching. In this form of construction, the span is composed of a series of beams with wide bottom flanges. A barrel vaulting built in brick carries the load from above to the bottom flanges of the beams, which are

M. H. GOULD

restrained against sideways displacement by rods running from side to side.

White Bridge

This form of bridge was never as common in Ireland as it was in Britain. The vast majority of those which were built have been replaced as being unsuitable for modern traffic loadings. It is therefore unusual to find two extant bridges of this type situated relatively close together.

White Bridge was built by Courtney, Stephens & Co. of Dublin, most likely in the 1860s. It has three longitudinal cast-iron I beams 12 in. by 24 in. Brick jack-arches span to these, although they were strengthened in 1985 by the application of one inch of gunite. The road is some 14 ft wide between the ornate iron balustrades. The bridge has four spans, each of 32 ft. It is supported on cylindrical columns with only a light cross bracing near the underside of the deck, which is now covered by lightweight concrete.

HEW 2141
J 216 380

Seafin Bridge was erected in 1878 by R. Lucus & Sons, Soho Foundry, Newry. It has three spans of 20 ft. On this structure, the brick jack-arching is in its original condition.

HEW 2134
J 049 499

23. Drumlone Bridge

HEW 2094
H 333 424

Drumlone Bridge, near Lisnakea in County Fermanagh, is the oldest reinforced concrete bridge in Ulster and is probably the second of its type to be erected in Ireland. It was built in 1909, using the Hennebique system, to a design by Mouchel & Partners. The span is about 42 ft 6 in. The design is typical of Hennebique bridges of this date and span: two deep longitudinal beams, a series of smaller cross beams above supporting the deck and cantilevered out to either side, and gas-tubing handrails. A few of these early concrete bridges are new decks on old piers (replacement of timber decks), but at Drumlone the abutments have also been rebuilt, at least from river level. Thus this probably replaced a masonry arch. The contractor was J. and R. Thompson. The bridge is a listed structure and was gunite coated in 1965 to improve the depth of concrete cover.

DE VESIAN J. S. E. Ferroconcrete in road bridge construction. *Proceedings of the First Irish Road Congress*, Dublin, 1910, 137.

24. Cruit Island Footbridge

HEW 3150
B 738 186

Cruit Island
Footbridge

This reinforced concrete footbridge, which connects Cruit Island to the mainland in County Donegal, was

DONEGAL COUNTY COUNCIL

built for the Congested Districts Board in 1911. It was designed by Mouchel & Partners and constructed using the Hennebique system of reinforcement.

The bridge has an arch slab 10 in. thick, being 7 ft wide at the springing narrowing to 4 ft 9 in. at the crown. The rise is 17 ft and the span some 90 ft. The deck spans about 110 ft between masonry abutments (which previously carried a timber deck) and is supported off the arch by three vertical slabs on each side.

This is an interesting early example of reinforced concrete bridge work and it is a pity that at the time of writing it is out of use and is being allowed to decay.

25. Silent Valley and Ben Crom Reservoirs

Once the Woodburn and Stoneyford catchments had been developed, it was considered by the 1890s that all local sources of fresh water at a level sufficient to serve the higher parts of Belfast had been utilised.

In 1891, the Belfast and District Water Commissioners therefore asked L. L. Macassey to consider what sources might be found for additional supplies. He reported that it would be necessary to develop a scheme in the Mourne Mountains. Accordingly, the catchments of the Kilkeel and Annalong Rivers were taken by an Act of 1893. A 40 mile long pipeline (including two long tunnels in rock) was constructed to deliver water to a reservoir at Knockbracken (J 365 663), flow commencing on 2 October 1901. The Commissioners also defined their catchment boundary by constructing the so-called 'Mourne Wall'. This 22 mile long rubble wall was built between 1904 and 1922.

HEW 1869
Sq J 32

Completion of the pipeline satisfied immediate needs and allowed the deferral of the construction of a reservoir dam until 1922, when S. Pearson & Sons were awarded the contract for the building of the Silent Valley Dam, designed to store 3000 million gallons. This is an earth embankment 1500 ft long and 88 ft high, with a clay core and concrete cut-off trench.

The Silent Valley Dam has become a *cause célèbre* for inadequate site investigation. Boring indicated rock at about 20–50 ft, but it was not proved. The contractor

HEW 1889
J 306 230

started to dig the cut-off trench, but found no rock, only boulders (one of which was described as being as big as a cottage), interlaced with soft clay. Rock was eventually found at 140–200 ft below ground level.

In 1928, the contractor was permitted to proceed on a cost plus basis, using a design which relied on the use of compressed air. A series of shafts with a diameter of 11 ft 10 in. was sunk. These shafts were filled with concrete to form the cut-off wall. A portion of the shafting, set up as a demonstration, still exists on site. Construction of the embankment itself was straightforward and the dam was opened on 24 May 1933. One highly attractive feature is the concrete bellmouth overflow, installed only after extensive model testing.

HEW 2119
J 317 260

In the valley above is the Ben Crom Reservoir, built 1953–57 by Charles Brand & Co. In terms of total height from foundation level, this is the second highest dam in Ulster, but at 125 ft it has the greatest free height above the river bed. It is a mass concrete gravity structure. This is a relatively uncommon form of construction in Ulster as earth embankments have usually been preferred. It was designed by Binnie, Deacon and Gourley.

LOUDEN J. *In search of water*. W. H. Mullan & Sons, Belfast, 1940, 82–149.

ANON. Silent Valley reservoir. *The Engineer*, 1931, **152**, 118.

ANON. Ben Crom reservoir. *Water & Water Engineering*, 1955, **59**, 191–196 and 1957, **61**, 446.

26. Spelga Dam Siphons

HEW 2093
J 266 273

The Spelga Dam, completed in 1957, is a conventional concrete gravity dam, 98 ft high by 52 ft wide (at base). The construction of the dam was straightforward, although some seams of sand in the underlying greywacke rock (due to igneous action) required careful grouting. The dam was designed to hold 600 million gallons of water. By 1971 it was becoming clear that additional storage would be needed to meet projected future requirements and various proposals were considered to meet this need. Eventually it was decided to increase the storage in the existing reservoir by 22 per cent by constructing siphons on the dam crest. These had to be designed to pass the maximum flood under the reduced

DAVID GAWN

head conditions resulting from the raising of the water level.

Spelga Dam
Siphons

The decision was made to use air-regulated siphons as producing the most stable conditions. Such a system had been installed at Eye Brook Reservoir, Leicester, in 1955, although later examples had been low head (e.g. for river flows). Accordingly, extensive laboratory studies were undertaken at the Queen's University of Belfast to confirm the design.

A major problem which had to be overcome at the Spelga Dam was that the siphons were required to saddle the existing crest, and this determined their basic shape. The siphons operate under a range of conditions from weir flow through partially primed conditions to full siphon flow.

Four banks of three siphons were constructed in 1974–75, two banks being either side of a vertical steel roller gate mounted on the existing overflow spillway. The siphons gave the required increase in storage capacity at a remarkably low cost. They are spectacular when in full siphon operation.

Design was by Ferguson and McIlveen of Belfast and the contractor was Charles Brand & Co.

POSKETT F. F. and SAYE J. A. The design and construction of the Spelga dam. *Proc. Instn Civ. Engrs*, 1959, **13**, 215–217.

POSKETT F. F. and ELSAWY E. M. Air regulated siphon spillways at Spelga dam. *Journ. Instn Water Engrs*, 1976, **30**, 177–190.

27. Dungonnell Dam

HEW 1854
D 192 171

Although of relatively recent construction, Dungonnell Dam is important because it was the first dam built in either Britain or Ireland which was waterproofed using an asphaltic lining laid on the upstream face. Four such dams had been built previously in Algeria, one elsewhere in Africa and one in Norway.

The dam itself is of straightforward design and construction with regard to cut-off and overflow arrangements and the resulting reservoir provides a water supply for Ballymena.The dam is built of general rockfill laid to slopes of 1 in 1.5 downstream and 1 in 1.7 upstream. It is some 950 ft long with a maximum height of 52 ft. It has a concrete cut-off wall and grout curtain near to the front toe. The upstream face consists of layers, 6 in. of small stone, 3 in. asphalt concrete underseal, and 5 in. open bitmac drainage layer, all sealed by 4 in. of asphaltic concrete.

Although the contractor had some problems in the design of equipment to lay the waterproofing to the required slope, these were overcome and the work proceeded without difficulty. The main contractor was John Rainey (Construction), Portrush; grouting was undertaken by Terresearch Ltd, London; waterproofing was by Wimpey Asphalt of London and the overall design was by Ferguson and McIlveen of Belfast.

POSKITT F. F. The construction of Dungonnell reservoir. *Paper to NI Association of the Institution of Civil Engineers*, 1971–72.

28. Altnaheglish Dam Stabilization

The dam at Altnaheglish near Dungiven in County Londonderry is of conventional mass concrete construction. When it was built in 1934 it was one of the highest dams in either Britain or Ireland. The main interest, however, centres on the problems associated with the concrete, which were first noticed in 1952 as spalling of the downstream face. Coring of the dam and abutments was undertaken, which revealed that little arch action to the abutments was possible as the rock had insufficient rigidity. Pressure grouting of the concrete and foundations undertaken in 1964 had no effect on the high water pressures in the rock and, accordingly, the 77 ft long central spillway crest was lowered in 1968 by some 6 ft. This reduced the capacity from 500 million to 390 million gallons.

HEW 1890
C 696 041

The increasing demand for water in the 1980s led to a re-examination of the dam. No chemical attack on the concrete was found, but the gneiss aggregate which had been used was found to have large expansion movements when subjected to cyclical wetting and drying. It was also calculated that the dam had a zero or negative factor of safety under certain conditions.

In order to restore full storage, a new overflow was constructed beside the dam and the old overflow sill was built up. A rock berm was placed in front of the concrete dam, and these two elements, when taken together, provided an adequate factor of safety. The original dam had valves in an internal gallery, protected by penstocks on the face of the dam, and special steps had to be taken to protect these. The remedial works cost £1.6 million and were undertaken by B. Mullan & Sons. The consultants were Ferguson and McIlveen of Belfast.

CRISWELL W. The Altnaheglish dam, Londonderry. *Conc. & Constr. Eng.*, 1933, **28**, 381–390.

COOPER G. A. Reservoir safety programme in Northern Ireland. *Journ. Instn Water Engrs*, 1988, **42**, 39–51.

29. Erne Drainage and Hydroelectric Development

HEW 3243
G 873 612 to
G 930 605

The 59 mile long River Erne, the second largest river system in Ireland, drains 1525 sq. miles of south-west Ulster. The river flows through Upper and Lower Lough Erne, discharging at some 164 ft above mean sea level at Belleek, and then dropping quickly by a series of rapids to the sea. The catchment regularly flooded because of the restricted discharge of the channel at Belleek, where the river runs on rock.

In 1863, the Lough and River Erne Drainage District Board was set up and constructed four flow control sluices at Belleek. The first drainage scheme undertaken by the Board was in 1885–90, but was restricted by an inability to excavate the rock at Belleek.

In 1918, the Board of Trade established the Water Power Resources Committee. Under the chairmanship of Sir John Purser Griffith, this body reported in 1921 on the potential for power generation in Ireland. In the late 1920s, work commenced on the Shannon Scheme, and it was not until 1945 that an Act was passed by the Irish government to allow work to commence at Ballyshan-

Overflow at
Cathaleen's Fall
Dam

M.H. GOULD

non. A corresponding Act was passed by the Northern Ireland government in 1947.

The works involved were the Cathaleen's Fall Dam, 844 ft long, 90 ft height, with an installed capacity of 45 MW; the Cliff Dam, 690 ft long, 60 ft high, with an installed capacity of 20 MW; the improvement of 3½ miles of river channel at Belleek, which involved the removal of 800 000 cu. yd of material, mainly rock, under a target price contract, and was undertaken at a cost 'considerably less' than the £1.125 million estimate; and the removal of the Belleek sluices and the construction of the Portora barrage at Enniskillen for water level control in Upper Lough Erne.

Construction work was undertaken by the Cementation Company Ltd and was completed by 1956.

ANON. Erne hydroelectric development. *Civil Engineering and Public Works Review*, 1948, **43**, 186–189.

PRICE J. Erne drainage. *Min. Proc. Instn Civ. Engrs*, 1889–90, **101**, 73–127.

30. Craigavon Bridge, Londonderry

Although the Swiss designer, John Conrad, built a 19 ft long model of a two-span timber bowstring bridge for the site in 1772, the first bridge across the River Foyle at Londonderry was constructed as a toll bridge at the relatively late date of 1791. It was superseded by the Carlisle Bridge, named after the Lord Lieutenant at the time, and opened by him on 25 September 1863. It was also a toll bridge until New Year's Day, 1878. The upper deck carried road traffic and the lower deck the mixed gauge railway lines (5 ft 3 in and 3 ft gauges) which linked the standard and narrow gauge stations on either side of the river. It was 30 ft wide, built of iron, and cost £90 000.

In 1933, the Carlisle Bridge was replaced by the Craigavon Bridge, which has a wider carriageway on the upper deck – 40 ft plus 10 ft footpaths. It again accommodated the mixed gauge railway goods lines belonging to the Londonderry Port and Harbour Commissioners on the lower deck. At either side of the river there was a turntable for the wagons, which were hauled across by ropes operated by capstans.

The bridge has five main spans, each of 130 ft, and 17

HEW 2127
C 437 161

approach spans, giving a total length of 1260 ft. Contemporary newspaper reports suggest that problems occurred in finding a firm foundation. Eventually, the cylindrical piers were sunk to a maximum depth of 69 ft below high water ordinary spring tides. The main Pratt truss type girders are spaced 36 ft apart. The Northern Ireland government provided £251 000 of the total cost of £255 500, leaving Londonderry Corporation to pay the balance.

The bridge was designed by Mott, Hay and Anderson and built by Dorman Long & Co. Ltd.

Work commenced on 13 April 1931. The bridge required 9100 cu. yd of excavation, 8600 cu. yd of concrete, and 5000 tons of steel. It was opened on 18 July 1933.

In 1968, after the closure of the Great Northern Railway track on Foyle Road in 1965, the lower deck was converted to a two-lane roadway. In 1984, the traffic congestion on the Craigavon Bridge was relieved by the opening of the high level Foyle Bridge some 2 miles downstream.

ANON. Londonderry's new bridge. *Civil Engineering*, 1933, **28**, 285.

31. Mullan Cot Footbridge

HEW 2126
H 946 806

The surprisingly modern looking concrete footbridge at Mullan Cot was built in 1924–25 to provide pedestrian access across the Ballinderry river to the chapel at Killymuck. It was funded jointly by the county councils of Tyrone and Londonderry as it spans the county border, although it was never adopted and has thus not been maintained at public expense.

The deck is carried on simple edge beams. These are supported on four pairs of legs without cross bracing. There are three main spans, each of 50 ft. The two supporting piers are located in the river, one being solid, the other consisting of two cylindrical piers, perhaps reflecting different foundation conditions.

The bridge was designed by Mouchel and Partners in Hennebique concrete and is believed to have been built by R. Calhoun Ltd of Londonderry.

The very simple lines of the design give the bridge a

M. H. GOULD

pleasing appearance and funds are being sought to incorporate it into a path for use by fishermen.

Mullan Cot
Footbridge

32. Burntollet New Bridge

Burntollet New Bridge was built in 1955 for Londonderry
County Council as part of the upgrading of the main road
between Belfast and Londonderry. The bridge used a
novel system of prestressing known as the Gifford-Udall-
CCL system. This was first introduced in 1953 and used
for a number of footbridges in Britain, including one in
St James's Park, London. The system was designed to
introduce higher levels of prestress at mid-span (the
point of maximum bending) than with conventional
prestressing. It is, however, more complicated to apply
and was used on only a small number of bridges. Burntollet is the only one of its type in Ireland and was the first
bridge to use the system when designed for M.O.T. standard loadings.

A hollow, solid-ended concrete trough was first cast.
Near its base were set horizontal prestressing tendons. At
Burntollet, these were 14 in. wide by 15 in. deep and there
were 36 primary wires and 12 untensioned. When cured,
the wires were stressed on the Cable Covers Ltd (CCL)

HEW 2128
C 502 108

197

system. The now stressed trough formed the bottom portion of a deeper beam. Additional formwork was placed together with post-tensioning ducts laid to a catenary. At Burntollet there were three, parallel at mid-span, but rising to end anchors one over the other. The remainder of the beam was now cast and the remaining tendons threaded (twelve per duct at Burntollet) and tensioned. On top of the beams a deck slab, incorporating further ducting for transverse prestress, was cast.

Burntollet has a 75 ft skew span with an angle of 38°. There are eight beams, 36 in. deep, which are waisted above the precast base for one-third of their length. The deck is 7 in. thick and is 40 ft between parapets. The contractor was Grahams of Dromore, County Down.

ANON. *The Gifford-Udall-CCL handbook.* Fred Phillips and Partners, London, 1957.

33. The Dry Arch

HEW 2129
D 035 361

An interesting unaltered example of early grade separation, where the gradient of the Coleraine and Dervock to Ballycastle road has been eased by passing it under the Ballantoy to Ballymoney and Armoy road.

The Dry Arch is a difficult bridge to date. A similar (but subsequently substantially rebuilt) bridge was formed in Banbridge (County Down) in 1831–34 when the main street was lowered into a cutting in order to ease the slope for the Belfast to Dublin mail coaches. There is, however, a note in the annual reports of the Board of Public Works (Ireland) that a loan was given in 1844 to improve the road between Ballycastle and Coleraine/Ballymoney and this was probably the source of funding for the construction of the Dry Arch.

The two roads were built to the standard pattern of Grand Jury roads of the mid-nineteenth century, the one crossing the other by a single span built of dressed stone.

34. Carnlough Harbour

HEW 1631
D 288 180

A pier was built at Carnlough in about 1800 by Philip Gibbons. Constructed using the method of dry stone walling, it was 200 ft by 20 ft wide. Limestone was trans-

ported by cart over a dry hurry (see Ducart's Canal) to the pier and by lighter to vessels lying offshore. The limestone was needed for iron smelting in Scotland.

Carnlough Harbour, circa 1934

The Londonderry family, who owned Seaham, County Durham, inherited land at Carnlough in 1834 and the harbour was improved. Richard Wilson, agent from 1850, oversaw the work. A protected basin was to be built, the hurry extended to drop-shoots at the pier (the system of coal loading at Seaham) and upgraded to an inclined plane.

Work began in late 1853, but, as the excavation for the basin proceeded seaward, problems were encountered with water inflow and when an igneous dyke was struck. It was not until late in 1855 that the basin became operational, and then only for ships drawing up to 8 ft. A loading quay was built on the south side.

Positioning of the basin had taken no account of the longshore drift. The old Gibbons pier, retained and strengthened for storm protection, cut right across this movement and siltation began at once. In 1859 a Lough Neagh dredger removed a large amount of material and the rest of the dyke. After part of the 30 ft high south pier collapsed and was rebuilt in 1860, the basin proved to be a moderate success, provided that dredging was maintained.

The ¾ mile of mineral railway, built to the English gauge of 4 ft 8½ in., ran by horse and gravity until 1898, steam until 1952 and electric traction until its closure in 1965. A 1½ mile branch, with a gauge of 3 ft 6 in., ran from 1890 until 1922. Ammonium sulphate (extracted from peat) was exported from 1902 to 1908 and processed lime from 1855.

The inclined plane required a 200 yd cutting (with the removal of 25 000 cu. yd of earth and rock) and bridges across the back and main streets. Built in white limestone, that at the main street bears a plate reading 'Carnlough Railway and Harbour. Projected and commenced by Charles William Vane, Marquis of Londonderry, 1853. Completed by Frances Anne Vane, Marchioness of Londonderry, 1854.'

The harbour is now used by pleasure craft.

IRVINE J. Carnlough harbour development scheme 1854–64. *Journ. of the Glens of Antrim Hist. Soc. (The Glynns)*, 1977, **5**, 22–30.

35. Gallaher's Factory Roof, Lisnafillan

HEW 1971
D 074 024

The main civil engineering interest at Gallaher's factory lies in the two main factory units, each of which has 20 curved post-tensioned roof beams of large proportions, giving clear spans of 101 ft. One unit has been lengthened by a further ten roof beams. Each beam is formed of precast concrete channel units. These were set upright on temporary staging to form a box section. The end units had cast-in anchors. Each box section (except at the ends) was separated by a ring section designed to hold the tendons to profile.

Once in place, multi-stranded tendons were fed down the inside of each beam and through the ring sections and anchors. Each strand was then tensioned and anchored (the CCL prestressing system), and the wires then grouted.

The roof was completed with curved concrete shells, 3 in. thick, surface insulation and an aluminium weatherproof layer. Aluminium-framed windows were put in

BROWNE OF LARNE

to form north lights. When built in 1956–58 this was the largest clear-span post-tensioned roof in either Britain or Ireland. The hollow centres of the beams, being 2 ft 8 in. wide by 7 ft 9 in. high, are used alternately for service pipes and air ventilation. Clean air and surfaces are important in cigarette manufacture and there was widespread use of precast concrete elements throughout the factory. The canteen, in particular, has eight curved concrete portal frames with 62 ft clear span.

Gallaher's factory roof, Lisnafillan

Design was by Sir Alexander Gibb & Partners. The main contractor was Sir Alfred McAlpine & Sons, with some work being undertaken by F. B. McKee & Co.

36. Concrete Bridges of Fermanagh

The partition of Ireland in 1921 affected the road pattern in County Fermanagh and led directly to the construction of a series of concrete bridges, most of which are multi-span.

HEW 2123 and HEW 2124

In Lower Lough Erne the main road was rebuilt across Boa Island, replacing ferry connections and keeping the road within Northern Ireland. The number of spans was reduced by the construction of long approach embankments. The bridge at Galloon Island in the Upper Lough also replaced a ferry.

M. H. GOULD

Lady Craigavon
Bridge, County
Fermanagh

By 1930, it was clear that those people living on the southern shore of Upper Lough Erne were having trouble reaching a shopping centre without crossing into the Irish Free State. Accordingly, Lisnaskea Rural District Council promoted the construction of a road across Trasna Island, Lisnaskea being a market town. Equal contributions to the cost of construction also came from Fermanagh County Council and the Free State government.

Table 4 : Fermanagh concrete bridges

Bridge	Grid ref.	Date of construction	Number of spans
1. Portinode	H 138 643	1925-27	3
2. Inishkeeragh	H 068 623	1925-27	3
3. Galloon Island	H 394 226	1928-31	3
4. Rosscor	G 987 586	1924-27	9
5. Lady Brooke	H 320 272	1931-35	13
6. Lady Craigavon	H 330 279	1931-35	11

In Table 4, bridges numbered 1 to 3 were designed and built by the Trussed Steel Concrete Company. They have an unusual feature in that the piles are carried up to top water level and the piers start at this level without any capping beam. Rosscor Viaduct, built by the same company, is mid-range in terms of dimensions. The spans are 35 ft centre to centre and the road width is 16 ft. In this case, each pier has five 14 in. square piles. The outer two are raked at 1 in 8 to counter the effect of river flow and a capping beam has been used. Rosscor has been damaged by explosion and is due for replacement.

Bridges 5 and 6 were designed by Mouchel & Partners and constructed using the Hennebique reinforcement system, the contractor being A. E. Farr of London. The navigation spans exhibit Mouchel patent large circular piers.

At Carry Island (H 294 373) is a very elegant modern concrete bridge, built by the Northern Ireland Department of Finance in 1954–56 as part of the works undertaken for the Erne hydroelectric development. It replaced an earlier masonry bridge dating from 1843.

Belfast Newsletter, 27 Jan. 1927.

Northern Whig, 12 Feb. 1927.

37. Number 6 Hangar, Ballykelly

At Ballykelly, County Londonderry, is a very large cantilevered roofed aircraft hangar. Built during the Cold War, many details of its construction are now lost. It appears to have been designed by the Ministry of Public Works and built around 1966. The steelwork was designed, supplied and erected by a Teesside contractor.

HEW 2142
C 633 242

The main structural element is a 120 ft long cantilevered girder, supported at one end on a vertical A-frame. The foundations appear to be large diameter friction piles. Extra concrete kentledge was added subsequently to the back of the A-frame. According to a local report, this was to counter movement due to the strong tidal effect on the groundwater level.

The clear covered floor area, accessible all along one side, is 724 ft by 158 ft, and was only surpassed by Brize Norton in England (1045 ft by 215 ft). More recent aircraft

Number 6
Hangar, Ballykelly

M. H. GOULD/M.O.D.

hangars have been designed as space frames, not as cantilevers.

GOULD M. H. Number 6 hangar, Ballykelly, Northern Ireland. *PHEW Newsletter*, 1996, no. 69, 1–2.

38. Coleraine Bascule Bridge

HEW 2096
C 846 333

At Coleraine is a bascule bridge carrying a single track railway over the River Bann downstream of the town. The central span lifts up by rotation at one end, to allow

COLERAINE CHRONICLE

the passage of ships to Coleraine harbour. The bridge is operated locally, but only after a large mechanical pin is released by the signalman at the Coleraine box. The lifting span, which was the first counterpoised bascule lifting span in either Britain or Ireland, is still operational, although commercial traffic to the harbour has all but ceased. The 250 ton lifting span is balanced by a large concrete block on steel cables underneath one end (the Strauss underhung system).

Bascule type bridges are rare in Ireland and that at Coleraine is the only example in Ulster still working. Other movable bridges, for example over the Bann Navigation, are generally of the swivelling type.

The Coleraine bridge was designed in 1921 by Sir W. G. Armstrong, Whitworth & Co. The left bank approach is curved and consists of four fixed spans, each of about 77 ft. The straight right bank approach has five similar spans. The lifting span is 85 ft and the overall width approximately 17 ft. Rail level is about 22 ft above high water. The superstructure is composed of Dorman Long steel and the reinforced concrete piers are founded on precast concrete piles.

Coleraine
Bascule Bridge

205

39. Foyle Bridge

HEW 2131
C 436 178

The Foyle Bridge near Londonderry has the largest single span of any bridge in Ireland. It was built to relieve congestion on the nearby Craigavon Bridge, and was opened early in 1984. An initial design was undertaken by Mott Hay and Anderson, but, following the success of the Kessock Bridge in Scotland, design and build tenders were sought in 1977. Consultants Ove Arup & Partners were appointed to assess the five bids received, and in December 1979, a tender for £15.7 million was accepted. The contract went to a consortium comprised of Redpath Dorman Long, Grahams of Dromore and Freeman, Fox & Partners. The winning design consisted of two parallel steel box girders with a curved concrete box section approach viaduct. The total length of the crossing is 865.7 m.

The main bridge has side spans of 144.3 m and a central span of 233.6 m. The box sections rest on two concrete river piers. Each steel box girder was built in parts by Harland and Wolff, Belfast, and transported to the site by barge. Four side-span units, each 180 m long and weighing nearly 1000 tonnes, were first placed. The single (full width) central span was then lifted vertically in between.

Navigation requirement was for a clearance over mean high water springs of 32 m for a width of 130 m.

Foyle Bridge

GRAHAMS-RDL

This required a prestressed concrete approach viaduct to be built from the eastern bank, the spans resting on five single and two double piers. Prestressing was on the PSC Freyssinet K system, using multi-stranded tendons.

The bridge, the profile of which was approved by the Royal Fine Art Commission, accommodates dual 7.3 m carriageways and 1.8 m footways and a central reservation.

PRESCOTT T. A. N. *et al.* Foyle Bridge: its history and the strategy of the design-and-build concept. *Proc. Instn Civ. Engrs*, 1984, **76**, 351–361.

WEX B. P. *et al.* Foyle Bridge: design and tender in a design and build competition. *Proc. Instn Civ. Engrs*, 1984, **76**, 363–386.

QUINN W. N. Foyle Bridge: construction of foundations and viaduct. *Proc. Instn Civ. Engrs*, 1984, **76**, 387–409.

HUNTER I. E. and MCKEON M. E. Foyle Bridge: fabrication and construction of the main spans. *Proc. Instn Civ. Engrs*, 1984, **76**, 411–448.

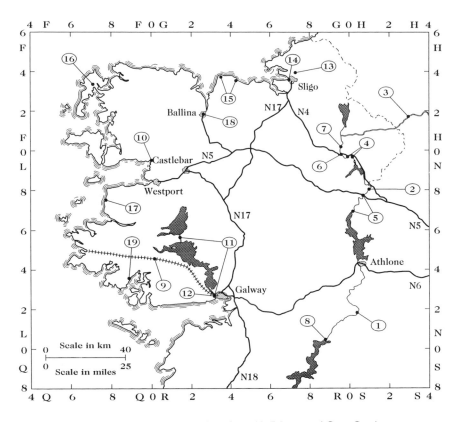

1. The Shannon Navigation (Killaloe to Athlone)
2. The Shannon Navigation (Athlone to Lough Allen)
3. Ballinamore and Ballyconnell Navigation
4. Drumsna and Jamestown Bridges
5. Lanesborough and Tarmonbarry Bridges
6. Carrick-on-Shannon Bridge
7. Hartley Bridge
8. Portumna Bridge
9. Galway to Clifden Railway
10. Newport Viaduct
11. Eglinton and Cong Canals
12. Salmon Weir Bridge
13. Doonally Reservoir
14. Hyde Bridge, Sligo
15. Dromore West and Rathlee Water Towers
16. Belmullet Water Tower
17. Louisburgh Clapper Bridge
18. Ballina Bridges
19. Flannery Bridge

6. Connaught

The western province of Connaught has a rugged Atlantic coastline stretching from Galway Bay in the south to the shores of Donegal Bay. The area of west County Galway and County Mayo is a complicated mixture of granite, quartzite and igneous rock intrusions. The limestone of the central plain reaches the coast in County Sligo. County Roscommon is mainly flat grasslands, the Shannon forming its eastern boundary, and the River Suck its border with County Galway. County Leitrim, by contrast, is characterized by lakes and hills.

The province includes some of the counties most affected by emigration and infrastructural development has been limited. Subsistence farming, fishing and tourism have been, until recent years, the main activities of the region. Important deposits of anthracite coal have been mined near Arigna in County Roscommon.

Only Galway and Sligo can claim to be established centres of manufacturing. Both are sea ports, Galway once being considered as a possible port for transatlantic liners. Sligo has a good example of a Victorian reservoir at Doonally, which still forms part of the town's water supply. It is, however, only within the last 20 years or so that many rural areas have received piped water on a regional basis.

There are no real barriers to communications, apart from the River Shannon on the eastern border of the province. Galway, Westport, Ballina and Sligo are served by rail from Dublin, but other lines have been closed, some as early as the 1930s, as they were totally uneconomic. The road network in the province has been considerably improved over the years.

Schemes for the improvement of the Shannon river system for the purposes of navigation were carried out in the second half of the eighteenth century and again in the first half of the nineteenth century. The latter scheme involved rebuilding or replacing many of the bridges across the waterway. A number of large locks and weirs were constructed and many shoals removed by dredging. An early 'clapper' bridge near Louisburgh in County Mayo is the finest example of its type in Ireland.

1. The Shannon Navigation (Killaloe to Athlone)

HEW 3074
R 705 730 to
N 039 416

In the late 1820s, the engineer John Grantham introduced steam navigation to the Shannon Navigation. Other promoters followed suit and steamer passenger and freight services became established, particularly between Athlone in County Westmeath and Killaloe in County Clare, but also as far north as Lanesborough and Carrick-on-Shannon.

The Grand Canal from Dublin terminates at Shannon Harbour, about halfway between Athlone and Portumna. The remainder of the navigation to Killaloe is through Lough Derg.

There are natural obstacles to navigation at four places, namely at Athlone, Shannonbridge, Banagher and at Meelick, north of Portumna. All the major engineering works required to make the river navigable have been associated with one or other of these locations.

In 1755, the engineer Thomas Omer began work for the Directors General of Inland Navigation and constructed a lateral canal with one lock to bypass the fall in the river at Meelick. The remains of this canal may be seen to the east of the present line of the navigation. Omer also supervised the construction of a canal at Athlone, the line of which may be seen to the west of the present cathedral. It was 1½ miles long and had one lock 120 ft long by 19 ft wide.

From the 1770s onward, the locks and associated engineering works fell into disrepair and a number of eminent engineers were consulted on how to remedy the situation, including Richard Evans, William Chapman and William Jessop. Following the Act of Union in 1800, the Directors General of Inland Navigation, acting on the advice of John Brownrigg, rebuilt the locks at Athlone and Meelick.

Shannon Commissioners were appointed in 1831 and extensive surveys of the navigation were carried out by their engineer, Thomas Rhodes. Under the terms of the Shannon Navigation Act of 1835, a combined drainage and navigation improvement scheme was undertaken in the 1840s, including steam bucket dredging of the main

R.C. COX

channel. At Meelick, the earlier canal was abandoned and a new large lock 142 ft long, capable of accommodating passenger steamers, was constructed in 1845 by the contractor William Mackenzie, and named Victoria Lock.

Victoria Lock, Meelick, on the Shannon Navigation

At Athlone, it was decided to deepen the river channel and to construct a substantial weir and lock (127 ft long) to replace the earlier bypass canal. The opportunity was also taken to build a new road bridge with an opening navigation span. Large caissons were used to allow construction in the dry of the bridge piers and the lock was built within a large cofferdam.

DELANY R. *Ireland's inland waterways*. Appletree Press, Belfast, 1992, 45–62, 123–134.

2. The Shannon Navigation (Athlone to Lough Allen)

The Upper Shannon Navigation extends northwards from Athlone, through Lough Ree, and continues north-westwards towards Carrick-on-Shannon and Lough Key. The navigation is also connected to Lough Allen and

HEW 3074
N 039 416 to
G 962 125

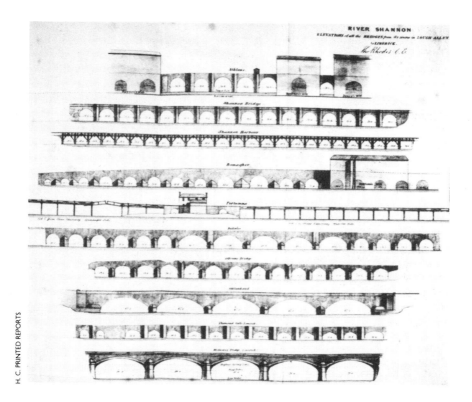

Shannon bridges survey, 1833

Lough Erne. The Royal Canal was opened to Richmond Harbour at Cloondara in County Longford by 1814. It enters the River Shannon near Tarmonbarry to the north of Lanesborough.

Apart from the reconstruction of a number of major road bridges, the main civil engineering work carried out along this stretch of the navigation was confined to the steam bucket dredging of new channels to bypass the earlier lateral canals constructed under Thomas Omer in the 1760s, the rebuilding of a number of locks and the deepening of the Jamestown Canal south of Carrick. This latter canal had been used by Omer to avoid shallows in the area by cutting across a natural loop in the river. The size of the locks gradually reduces as one moves north, an indication of the perceived commercial potential of the navigation.

The Shannon and Erne systems are now connected by

the totally rebuilt Ballinamore and Ballyconnell Navigation (now known as the Shannon–Erne Waterway) and represent an important tourism and recreational resource.

DELANY R. *Ireland's inland waterways*. Appletree Press, Belfast, 1992, 45–62, 123–134.

3. Ballinamore and Ballyconnell Navigation

Built with the aid of government funding under the Arterial Drainage Act of 1842, the Ballinamore and Ballyconnell Navigation (its original name) commences with a still-water canal of 7½ miles from the River Shannon near Leitrim village to Lough Scur, using a flight of eight locks in a 3 mile stretch. The navigation then comprises a summit level channel leading from Lough Scur to Lough Marrave, a distance of 3 miles, the remainder of the route to Lough Erne being by way of the canalized Woodford and Yellow rivers. The total length of the navigation is 38½ miles.

HEW 3137
G 957 046 to
H 353 218

The original navigation was designed by William Thomas Mulvany (1845) and John McMahon (1846) and was built by direct labour. Mulvany was a Drainage Commissioner who, on leaving the Board of Works, went to Germany where he, with other partners, developed the rich coal deposits, so laying the foundations of the Ruhr economy.

The navigation was opened in 1860, but was never a commercial success and was closed by 1869.

In 1990 work commenced on an ambitious and expensive project to connect the Shannon and Erne waterways to enhance their tourism and recreational potential. This was a joint venture between the Irish and United Kingdom governments and was also supported by the Ireland Fund and other agencies. The work included the complete reconstruction of all the locks, those at the Leitrim end being 82 ft long by 16 ft 3 in. wide with a water depth of 5 ft 6 in. The locks between Lough Scur and Lough Erne are somewhat wider. All masonry over-bridges were either restored or rebuilt, depending on their condition.

New steel gates, with timber heel and mitre posts and bottom rails, were fitted to all 16 locks.

The navigation, with its new name of the Shannon–Erne Waterway, was reopened to traffic on 2 April 1994.

FLANAGAN P. *The Shannon–Erne waterway*. Wolfhound Press, Dublin, 1994.

4. Drumsna and Jamestown Bridges

Until recently, the main road from Dublin to Sligo crossed a loop of the River Shannon in County Leitrim at two places, Jamestown and Drumsna. This reach of the river was never part of the main navigation, as in the 1760s Thomas Omer cut a 1.3 mile long canal across the loop to the south. A substantial lock was added in the 1850s at the eastern entrance to the canal from the Shannon and the so-called Jamestown Canal (HEW 3194) was straightened.

HEW 3147
M 993 971

The older of the two road bridges, that at Drumsna, has seven segmental masonry arches, increasing in span from 20 ft at the west end to 29 ft at the centre of the river, again reducing to 23 ft at the east end. Another 20 ft span arch at the east end was subsequently blocked off. More recently, the ring of the most westerly arch was replaced and the soffit concreted. Several tie bars and plates prevent the spandrel walls from spreading outwards. The pointed cutwaters are taken up to parapet level to form pedestrian refuges. These were deemed to be necessary as the bridge is only 16 ft between the parapets. Drumsna Bridge appears to have been erected in the early years of the eighteenth century and is largely unaltered.

HEW 3191
M 981 970

Jamestown Bridge was completed in 1847 by the Shannon Commissioners under their engineer, Thomas Rhodes. It has five segmental arches of 30 ft span and 7 ft rise. The large voussoirs in the arch rings are of ashlar limestone, as are the spandrel walls. The cutwaters are rounded, rather than the pointed examples more typical of earlier bridges. On the same road alignment as Drumsna Bridge, the bridge at Jamestown was designed to be of similar width.

The national primary road now bypasses this part of the Shannon, thus ensuring that the bridges at both

R. C. COX

Drumsna and Jamestown will continue their important
contribution to the pleasing environment of the Upper
Shannon.

Drumsna Bridge

5. Lanesborough and Tarmonbarry Bridges

Following the passing of the Shannon Navigation Act in
1835 and the appointment of the Shannon Commissioners, work on improving the navigation between Killaloe
and Carrick-on-Shannon commenced in 1840. The numerous road crossings of the river were the subject of
major engineering works, and in many instances completely new bridges replaced a variety of structures, some
timber, some masonry.

At Lanesborough in County Longford, to replace an
eight-span masonry bridge, one of six 30 ft spans, with a
navigation opening of 40 ft spanned by an swivel bridge,
was completed in 1843.

HEW 3188
N 006 695

215

The swivel bridge was removed in the 1970s and re-placed with a fixed reinforced concrete beam and slab arrangement allowing sufficient headroom in the navigation channel.

In 1993, a reinforced concrete deck with cantilevered footpaths was built on top of the earlier masonry structure in order to provide the necessary width for the national primary road between Longford and Roscommon.

HEW 3190
N 055 770

At Tarmonbarry, the Shannon is divided into two channels by an island, 85 ft wide along the centre line of the bridge. The crossing is composed of two bridges, one of three segmental arches over the east channel and one of four over the west channel, all of 33 ft span with a rise of 8 ft. The navigation opening on the west side of the river, like that at Lanesborough, was originally spanned by a swivel bridge. This was replaced in the 1970s by the present 40 ft span vertical lifting bridge. The piers are 6 ft 6 in. thick, except those either side of the navigation opening, which are 20 ft. The Tarmonbarry bridges replaced an earlier seven-span masonry arch bridge whose foundations were found to be poor and not suitable for underpinning. A reinforced concrete deck with cantilevered footpaths was added in 1993 to accommodate the main road from Longford to Strokestown.

6. Carrick-on-Shannon Bridge

HEW 3146
M 938 993

During the 1840s, the improvement of the Shannon Navigation required many of the bridges spanning the river to be rebuilt or replaced. Passenger steamers operated between Athlone and Killaloe, but there was less traffic at the northern end of the navigation and the locks were smaller. Opening spans were normally incorporated into bridge crossings, but by the time the improvement works had reached Carrick-on-Shannon, money was running low and it was decided to construct a new fixed bridge at this point with no opening span.

The bridge, designed by Thomas Rhodes, was completed in 1847 and has five spans built of ashlar limestone. These increase from 30 ft to 35 ft towards the centre of the bridge, with a rise to span ratio of 0.23, giving a gradual

R. C. COX

slope to the parapet. The roadway, however, is level over the central part of the bridge with ramps at each end. There is a smaller arch in the eastern abutment which allows pedestrian movements along the quays. The contractor for the bridge was William Mackenzie.

Carrick-on-Shannon Bridge

7. Hartley Bridge

Financed jointly by Roscommon and Leitrim County Councils and the Board of Works, Hartley Bridge was opened in 1915. It carries a minor road over the River Shannon between Carrick-on-Shannon and Cootehall.

HEW 3187
G 938 022

An early example of the use of reinforced concrete in Ireland, the bridge has 48 in. deep parapets which act as beams spanning between supports. The supports consist of pairs of 12 in. by 12 in. square columns with cross ties and diagonal bracing struts. The main reinforcement used was of a rail type section, known as a Moss Bar.

The spans from centre to centre of the supports are

217

R. C. COX

Hartley Bridge (from the west end of the bridge) three of 36 ft, one of 40 ft, and one over the navigation channel of 60 ft.

The design was prepared for Eugene O'Neill Clarke, County Surveyor of Leitrim and the work was carried out by direct labour.

BARRY M. B. *Across deep waters: the bridges of Ireland*. Frankfort Press, Dublin, 1985, 133.

8. Portumna Bridge

HEW 3161
M 871 043

A timber trestle bridge was erected across the River Shannon at Portumna in 1795 by the American engineer Lemuel Cox. This was one of seven similar bridges erected in Ireland by Cox. The Portumna Bridge was partially rebuilt in 1818 and replaced in 1834 by another timber structure incorporating an iron swivel bridge over the navigation channel on the Galway side of the river, and causeways forming the approaches to the bridge. This bridge was in turn replaced by the present structure, which was completed in 1911.

The Shannon at this point consists of two channels divided by Hayes Island, the one on the North Tipperary side being about 260 ft wide, and that on the Galway side being about 240 ft wide. Each channel is spanned by three pairs of mild-steel plate girders (either 80 ft or 90 ft in length) resting on 9 ft diameter concrete-filled cast-iron cylinders. The width of the approach roadways and bridge is 30 ft. The swing bridge over the 40 ft wide navigation channel has unequal arms of 60 ft and 30 ft length respectively and revolves on a pivotal support on the Galway bank of the river. The river piers are continued upwards beyond parapet level, tapering to domed tops with decorated finials. The earlier substantial masonry abutments were retained when the replacement bridge was erected.

The bridge was designed by C. E. Stanier of London to the specification of J. O. Moynan, the County Surveyor of Tipperary (North Riding). The contractors were Hernan and Froude of the Newton Heath Ironworks in Manchester.

MOYNAN J. O. A short description of the existing bridges at Waterford and Portumna and of the proposed new structures to replace them. *Trans. Instn Civ. Engrs Ir.*, 1910, **36**, 224–250.

Portumna Bridge

R. C. COX

9. Galway to Clifden Railway

HEW 3256
M 303 254 to
L 658 507

Under the Light Railways (Ireland) Act of 1889, the Midland Great Western Railway Company (MGWR) was provided with a government grant of £264 000 to build a line from their terminus at Galway across the rugged grandeur of Connemara to Clifden. The intention had been to improve communications with a developing fishing industry and the MGWR engineers designed a route to follow the coastline, where the population was estimated to be around 60 000 persons. However, the Royal Commission on Public Works thought otherwise and directed that an inland route should be followed via Oughterard. Largely as a result of this decision, freight traffic failed to materialize, and the railway company chose instead to develop the tourism potential of the area.

The total length of the single-track line was just over 48 miles and was built to the Irish standard gauge of 5 ft 3 in. The line was opened to Clifden on 1 July 1895 and cost £432 000, or £9000 per mile. Altogether, there were some 30 bridges, including an imposing steel viaduct, which crossed the River Corrib in Galway. Today only the piers remain of the viaduct, the three spans of which were 150 ft, with a bascule type lifting navigation span of 21 ft.

The engineers for the railway were John Henry Ryan and Edward Townsend. Robert Worthington won a provisional contract to build the line, but the final contract went to a Charles Braddock. However, being described as a 'mere adventurer and pauper contractor', he was eventually replaced by Travers H. Falkiner. The line was closed in April 1935.

RYAN J. H. The Galway and Clifden Railway. *Trans. Instn Civ. Engrs Ir.*, 1902, **28**, 203–235.

VILLIERS-TUTHILL K. The Connemara Railway 1895–1935. *History Ireland*, 1995, Winter, 35–40.

10. Newport Viaduct

HEW 3107
L 983 939

The Light Railways (Ireland) Act of 1889 specifically targeted districts where the development of fishing or other industries would benefit and where circumstances of distress in those districts necessitated state aid. Under

R. C. COX

the Act, the Midland Great Western Railway Company (MGWR) constructed branches to Clifden in County Galway and to Achill and Killala in County Mayo.

Newport Rail Viaduct and Road Bridge

The town of Newport is on the Newport river, where it flows into the north-east corner of Clew Bay. In late 1890, Robert Worthington, the main contractor to the MGWR, began work on the line between Westport and Mallaranny (later extended to Achill). Design was by the railway company's engineer, William Barrington.

At Newport the single line track crossed the river on a fine viaduct of red sandstone. It has seven segmental arches of 37 ft span with a rise of 12 ft 6 in., the arch rings being 24 in. thick. The overall length of the viaduct is 305 ft and the width 18 ft 6 in. The piers are 6 ft thick with pointed cutwaters.

The date 1892 is carved on the parapet, although the line was not opened until 1894 on completion of a nearby tunnel. The viaduct is now maintained as a pedestrian route and, together with the adjacent road bridge, forms an attractive backdrop to the town, especially when floodlit at night.

ROWLEDGE J. W. P. *A regional history of railways, Volume 16: Ireland.* Atlantic Transport Publishers, Penryn, Cornwall, 1995, 163–164.

11. Eglinton and Cong Canals

HEW 3090

Lough Corrib in County Galway, the second largest lake in Ireland, was used from earliest times as a trading route. Improved access to the city of Galway, and even to the sea, had long been the ambition of promoters of various navigation schemes.

As early as 1178, the friars of Claregalway Abbey had dug an artificial cut through an island in the River Corrib to shorten the distance to the city. This was later widened and became the main navigation channel. In 1498, an attempt was made by Andrew Lynch to cut a canal to the east of the city through to Lough Atalia, but this remained unfinished.

M 296 247 to
M 293 256

Parkavera Lock,
on the Eglinton
Canal

A basin for shipping at the Claddagh was completed in 1830 to a plan of Alexander Nimmo and John Killaly, but it was not until 1848 that construction of the Eglinton Canal linking Lough Corrib with the sea began. Designed by John McMahon, it was built by direct labour with a grant from the Board of Works. The canal, which could

R.C. COX

accommodate vessels up to 125 ft in length and of 20 ft beam, is about three-quarters of a mile long, with two locks, and was opened in 1852. Low bridges were built across the canal in the 1950s, thereby obstructing the navigation.

Work also began in 1848 on the Cong Canal to link Lough Corrib with Lough Mask to the north, but swallow-holes in the limestone bedrock, and the decision by the Board of Works in the 1850s to limit the works to the regulation of drainage, resulted in the works being abandoned.

M 149 553 to M 126 595

Discounting a late eighteenth-century folly lighthouse, erected near Kells in County Meath, Ballycurrin Lighthouse (M 195 490), erected in 1775 on the eastern shore of Lough Corrib, is the only example in Ireland of an inland lighthouse used as an aid to marine navigation.

DELANY R. *Ireland's inland waterways*. Appletree Press, Belfast, 1992, 170–174.

12. Salmon Weir Bridge, Galway

The River Corrib, flowing out of Lough Corrib through the city of Galway, is spanned by a number of bridges, the best known being the masonry segmental road bridge immediately downstream of the salmon weir.

HEW 3120 M 295 255

The architect, Vitruvius Morrison, designed a classical structure in ashlar limestone, the rounded capped cutwaters and balustraded parapets being particularly attractive details.

The bridge has seven spans, varying from 29 ft 2 in. to 37 ft 2 in., two of which are over the so-called Eastern Conduit, an artificial leat or open watercourse, separated from the river by an embankment and previously supplying a number of mills downstream.

The piers and abutments of the bridge are founded on solid rock, which lies near the surface at a point where the river begins a more rapid descent to the sea.

Behan is thought to have been the contractor for the sponsors, the Corrib Drainage Commissioners, and the bridge was opened in October 1818.

It is said that the Salmon Weir Bridge was built at this

JOHN HINDE, FRPS

Salmon Weir
Bridge, Galway

location in order to provide a convenient route from the
courthouse on one side of the river to the jail on the other!

CANNON E. *Survey of bridges in Galway City*. Galway Corporation, Galway, 1990.

13. Doonally Reservoir

HEW 3008
G 727 395

The present water supply to Sligo town and its environs
is divided approximately equally between the supply
from Lough Gill to the south-east and that from the
Doonally Reservoir to the north-east of the town. The
latter, known as the Kilsellagh Water Supply Scheme, is
a typical example of late Victorian water engineering,
other examples being the works at Omagh in County
Tyrone and Rathmelton in County Donegal (both designed by the Dublin firm of Pallas and Walsh).

The water from the Doonally and Kilsellagh streams
is impounded by an earthen dam approximately 30 ft
high and 600 ft long and is led some 3 miles by piped
aqueduct to Sligo. A third stream, the Carrowlustia, is led
around the south side of the reservoir in a lined channel.
The other streams can be diverted around the reservoir

in times of flood by similar relief channels on the north side.

The inner face of the dam, with a puddled clay core to prevent seepage, is rubble faced at a slope of 1 in 3, whilst the outer face is grassed. At the southern end of the dam, a semicircular weir, 60 ft long, leads overflow water by a 9 ft wide byewash, with a segmental-shaped invert, into the Doonally Stream below the dam. The draw-off tower has three valve controlled openings, and is accessed from the roadway on top of the dam by a wrought-iron latticed girder footbridge of two 90 ft spans.

The scheme, completed in 1884, was designed by the Westminster consulting engineering firm of R. Hassard and A. W. Tyrell and the contractors were Sweeny and McLarnon.

ANON. Sligo Corporation waterworks contract 1: construction of reservoirs. *The Engineer*, 1881, **51**, 160–161.

14. Hyde Bridge, Sligo

Sligo occupies a site which has been of importance from earliest times. The River Garavogue, flowing out from Lough Gill to the east, narrows and falls over a series of rock ledges as it nears its estuary. The position of a strategic ford is marked by the older of two town bridges, the so-named New Bridge, a seventeenth-century multi-span rubble masonry bridge with semicircular arches.

HEW 3009
G 689 364

The second of Sligo's present bridges was erected between 1846 and 1852 on the site of an earlier twelfth-century timber bridge. This bridge was itself replaced in the fourteenth century by a narrow eight arch skewed masonry bridge. It was rebuilt many times before being demolished in 1846 to make way for the present bridge.

This is a five-span segmental arch bridge of ashlar limestone, with solid parapets and rounded cutwaters on the upstream face. The spans are all 21 ft 3 in. with a rise of 6 ft 10 in., resulting in a level parapet and roadway. Architectural detailing includes a projecting cornice with dentils and archivolts at the top of the arch rings. The piers and abutments are founded on a rock ledge over which the river flows into the tidal estuary.

The design was the work jointly of Sir John Benson and

R. C. COX

Hyde Bridge, Sligo — the County Surveyor of Sligo, Noblett St Leger. The bridge was opened in 1852 as Victoria Bridge, but was later renamed in honour of Ireland's first president, Douglas Hyde.

O'RORKE, T. *The history of Sligo: town and county.* Duffy & Co., Dublin, 1889, Volume 1, 335, 556–557.

15. Dromore West and Rathlee Water Towers

HEW 3265
G 429 358
G 340 377

Regional water supplies were slow to be provided to the less heavily populated areas of the west of Ireland. The flat coastal plain along the southern shores of Sligo Bay was supplied by the Lough Easkey scheme, completed in 1986.

The scheme called for two water towers of the more usual modern 'wineglass' type. The contractor, Uniform

Construction Co. of Dublin, suggested the use of patent Rund-Stahl-Bau shuttering from Germany. This was accepted and resulted in the first use of the system in either Britain or Ireland.

Both towers have a simple 4.3 m diameter shaft on a 12 m diameter pad base. The bowl of the tower is some 6.5 m in height and the sides slope at an angle of 45°.

The shaft was constructed in conventional short lifts. The patent shuttering was then erected and supported on the shaft, thus alleviating the need for the forest of scaffolding usually needed for such towers. The bowl was poured in one day without breaks, thus ensuring a clean finish without construction joints. Each tower was finished in some 30 weeks.

The consultant for the towers was Jennings O'Donovan & Partners and the shuttering was used subsequently for the construction of a number of other towers in the Republic of Ireland.

16. Belmullet Water Tower

Belmullet in County Mayo, like many small rural towns in Ireland, had only a small local water supply until recently. As steps were taken to provide a regional water supply, a new source was developed in 1985 from Carrowmore Lake.

HEW 3260
F 705 335

In order to ensure sufficient storage and head for the town and the Mullet peninsula, it became necessary to construct a water tower. The site at Tallagh was very exposed and considerable thought was given to the most appropriate design. The result is a fine example of a modern water tower, which was constructed in 1992.

The tower consists of three round tanks of different height and diameter, each on its own circular shaft. Each tank has a central access, with the overall access through the largest. The whole effect, with the vertical features on the walls, is reminiscent of a set of enmeshed cogwheels.

The design was prepared by Ryan Hanley & Co., Galway, and the tower was built by Uniform Construction Co. Ltd.

All three tanks have a common top water level and the total capacity is 2165 cu. m. The largest has an internal

Belmullet Water
Tower

L LYONS, FIPPA, WESTPORT

diameter of 17 m and is 6.7 m high; the next, with the same height, is 14 m in diameter; and the smallest is 10 m in diameter but is 8 m deep. The aspect ratio of the tanks is altered by the use of parapets above roof level. Maximum height is 26 m.

As with all regional water schemes, the construction of the tower was accompanied by the installation of a considerable network of water mains.

17. Louisburgh Clapper Bridge

HEW 3241
L 754 759

In southern England, particularly in Devon and Cornwall, there are a number of well-known clapper bridges. The word clapper comes from the Latin *claperium*, meaning a pile of stones. Such bridges were constructed for pedestrians across rivers at places were the water depth was small at times of normal flow and were often associated with an adjacent ford.

The largest complete clapper bridge in Ireland is that located at Burlehinch to the west of Louisburgh in County Mayo. It is around 164 ft long and carries a

R. C. COX

footpath across streams flowing into Roonagh Lough. Agricultural traffic on the associated minor road fords the streams alongside the bridge.

Louisburgh
Clapper Bridge

The bridge has over 30 spans, the clappers or flat limestone slabs varying in length from 2 ft 6 in. to 5 ft. The slabs are generally about 2 ft wide and rest on random rubble piers of average height 2 ft above bed level. There are 2 ft high parapet walls on the downstream side with 12 in. by 12 in. openings centred over each span. These openings allow the water to flow unimpeded during times of flood.

No records have been found relating to the construction of the Louisburgh bridge, but local historians believe it to be of 'great age', and it could date from medieval times. The remains of a small clapper bridge at Ballybeg, Buttevant, County Cork are believed to date from the foundation of a nearby abbey in 1229.

O'KEEFFE P. J. and SIMINGTON T. A. *Irish stone bridges: history and heritage.* Irish Academic Press, Dublin, 1991, 73.

OTTER R. A. (ed.) *Civil engineering heritage: Southern England.* Thomas Telford, London, 1994, 92–93.

229

18. Ballina Bridges

HEW 3197
G 246 188

The famous salmon fishing river, the Moy, is spanned at Ballina in County Mayo by two road bridges. The town of Ballina was the first to be captured by the little band of 1100 Frenchmen, under General Humbert, who landed at Killala Bay during the insurrection of 1798, so this is probably a long established crossing point.

The most westerly of these two bridges was designed, according to a plaque on the parapet of the bridge, by the civil engineer, Thomas Ham. He is known to have completed the lighthouse at Black Rock, Sligo in 1835.

The present bridge is a fine masonry structure reminiscent of many of the bridges erected by the Board of Works. It has five segmental arch spans, each of 35 ft with a rise of 8 ft, and was erected in 1836. The work is of high quality ashlar masonry with well-formed arch rings.

The bridge further downstream is of four arches and appears to be of a slightly earlier date, possibly about 1830.

19. Flannery Bridge

HEW 3244
L 867 354

Flannery Bridge spans a creek on the main Maam Cross to Carna road in County Galway. The original bridge, of stone pier and girder construction, was erected in 1887 at the behest of a local priest, Father Flannery. As the carriageway was only 6 ft 6 in. wide, it was replaced in 1954 by the present bridge, which consists of a single-span two-pin prestressed concrete arch with a clear span between concrete abutments of 171 ft. The bridge deck, which is simply reinforced, is supported on four arched ribs, the depth of the ribs reducing to 3 ft at the centre of the span. The ribs represent an early example of the use of the Freysinnet system and the first prestressed concrete statically indeterminate bridge structure to be built in Ireland. The first prestressed concrete bridge in Ireland is believed to have been a road bridge constructed in 1952 by Córas Iompair Éireann to carry the Naas to Rathangan road over the main Dublin–Cork railway line. The designers in this case used the Lee McCall system. The prestress in Flannery Bridge was applied to cables made up of twelve 0.276 in. diameter high tensile steel wires

Flannery Bridge

M. B. BARRY

with spiral core. The bridge is now closed and traffic is diverted across a temporary causeway, pending the construction of a new bridge.

O'LEARY D. and MURPHY C. Design and construction of a prestressed bridge. *Trans. Instn Civ. Engrs Ir.*, 1954–55, **81**, 125–137.

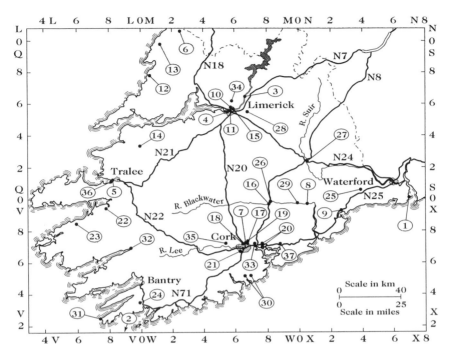

1. Dunmore East Harbour and Lighthouse
2. Fastnet Rock Lighthouse
3. Limerick and Killaloe Navigation
4. Limerick Docks
5. Tralee and Blennerville Ship Canal
6. Ballyvaughan Water Supply
7. Cork Water Works
8. Lismore Bridges
9. Causeway Bridge (Dungarvan)
10. Thomond Bridge
11. Sarsfield Bridge
12. Bealaclugga Bridge
13. Spectacle Bridge
14. Listowel Bridge
15. Athlunkard Bridge
16. Fermoy Bridge
17. St Patrick's Bridge, Cork
18. Daly's Bridge, Cork
19. Cork Rail Tunnel

20. Cork and Cobh Railway Viaducts
21. Chetwynd Viaduct
22. Laune Viaduct
23. Gleensk Viaduct
24. Ballydehob Viaduct
25. Mahon Viaduct
26. Carrickabrack Viaduct
27. Cahir Viaduct
28. Barrington Bridge
29. Ballyduff Upper Bridge
30. Eastern Bridges (Kinsale)
31. Mizen Head Bridge
32. Kenmare Bridge
33. Lee Road Tunnel
34. Shannon Power Scheme (Ardnacrusha)
35. Inishcarra Dam
36. Blennerville Windmill
37. Midleton Distillery Water Wheel

7. Munster

The province of Munster covers much of the south and south-west of Ireland. The geology of the region is mainly a system of east and west foldings of old red sandstone. Limestone generally occupies the valley bottoms between the upland folds, and in places the rivers have cut gorges through the ridges. To the west the ridges pile up into mountains, reaching their maximum height of over 3400 ft in County Kerry, the highest in Ireland.

The ports of Cork, Limerick and Waterford have each played an important part in the development of their associated hinterlands. Cork, the second largest city in the Republic, stands on the River Lee at the entrance to Lough Mahon and has a large land-locked natural harbour. Industries include brewing and distilling, steel making, hosiery and chemicals.

The Munster coastline can be hazardous to shipping. Lighthouses, such as that on the Fastnet Rock, have helped to warn shipping approaching from the Atlantic along one of the main shipping routes from North America to Europe.

County Limerick has a fertile limestone plain in its centre and ranges of hills on its borders. Limerick city lies at the head of the estuary of the River Shannon. The city has large wet docks and a number of fine bridges. Nearby, at Ardnacrusha in County Clare, is the works of the Shannon hydroelectric scheme, the first to be built in Ireland and a magnificent achievement by the engineers and contractors involved with the project.

Construction by the Great Southern and Western Railway Company of the main route between Dublin and Cork was completed by 1855, Limerick having been reached in 1848. Tralee was connected by 1859 and Waterford in 1853. The northern approach to the city of Cork required the construction of a number of substantial viaducts and the driving of a tunnel nearly one mile in length. The railway network was extended in the 1880 and 1890s to serve outlying areas such as Valentia, Kenmare and Dingle.

Clare is mostly flat limestone country and presents few obstacles to communication. The largest centre of commercial activity is now Shannon Airport, with its associated industrial estates.

Some well engineered mountain roads and bridges were built in the Cork/Kerry region in the 1820s and 1830s under the supervision of the government agent, Richard Griffith (Figure 5).

The main river systems in County Waterford are the Suir and the Blackwater. The Blackwater flows out to the sea at Youghal in County Cork, whilst the Suir forms the border with counties Tipperary and Kilkenny.

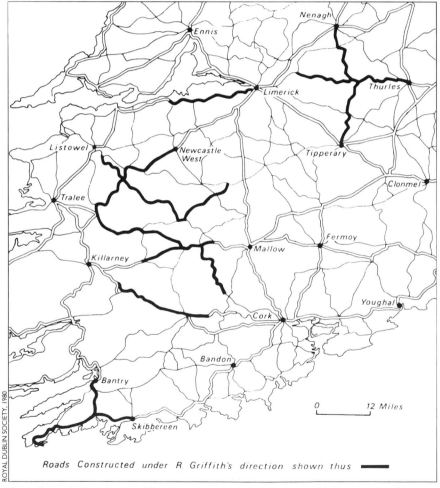

Figure 5. Roads constructed in Munster under the direction of Richard Griffith between 1822 and 1836

1. Dunmore East Harbour and Lighthouse

The original pier and lighthouse at Dunmore East on the south-east coast of County Waterford at the entrance to Waterford Harbour were built under the direction of the civil engineer Alexander Nimmo, between 1823 and 1825. The harbour was built for the British Post Office to accommodate the mail packet steamers operating out of Milford Haven in South Wales.

HEW 3072
S 690 000

Nimmo's Pier and Lighthouse at Dunmore East Harbour

R. C. COX

The pier is about 780 ft long and 77 ft wide at the base. A substantial parapet wall rising to a height of 14 ft above the surface of the pier provides shelter from easterly gales, the seaward side of which is protected by battered masonry slopes or pavement.

The pier terminates in a bull-nosed platform on which stands the lighthouse, a fluted Doric column of old red sandstone conglomerate 50 ft high.

The pier was partly rebuilt in 1832 by the Board of Works under the direction of Barry Duncan Gibbons and a further 16 ft wide breakwater, extending 295 ft long to the north-west from the end of the pier, was added in the 1960s. Also at this time, an extra storm wall extending for a length of 666 ft was constructed over the original pier parapet. It is 10 ft 4 in. wide at the crest, about 16 ft wide at the base, and incorporates a curved wave deflector. Between 1963 and 1972, Dunmore East Harbour was further developed as a major fishery port.

2. Fastnet Rock Lighthouse

HEW 3049
V 882 163

On a jagged pinnacle of rock, $4\frac{1}{2}$ miles off the Cork coast south-west of Cape Clear, stands the most famous of all Irish lighthouses. The first lighthouse tower on the Fastnet Rock, designed by George Halpin (Senior), the Engineer to the Ballast Board in Dublin, was constructed between 1848 and 1853. It replaced the light on Cape Clear, which was frequently shrouded in fog because it was 450 ft above sea level.

The cast-iron tower, supplied and erected by J. and R. Mallet of Dublin, was 63 ft 9 in. high, tapering from a diameter of 19 ft at the base to 13 ft 6 in. at the top. It consisted of flanged plates bolted together, the base being secured by long bolts into the rock surface. The base was filled solid with rubble masonry to a depth of 3 ft and lining walls of masonry 3 ft 6 in. thick were built up to first floor level. The lantern added another 27 ft 8 in. to the overall tower height of 91 ft, the light itself being 148 ft above sea level. The tower was modified in 1867 to make it more resistant to heavy sea conditions, but doubts were raised regarding the stability of such towers, and it was decided by the Commissioners of Irish Lights in 1891 to build a completely new lighthouse on the rock. The old tower was taken down as far as the room immediately above the solid base section, the room being converted into an oil storage tank for the new lighthouse.

Following a considerable amount of preparatory work, construction of the present tower, designed by William Douglass (Engineer-in-chief to the Commissioners), was commenced in 1896 and continued, as weather permitted, until June 1904, when the powerful light was exhibited for the first time. The tower is built from Cornish granite, supplied by John Freeman & Co. of Penryn, each of the granite blocks being dovetailed into adjacent blocks in each ring of masonry. The tower contains 58 000 cu. ft of masonry weighing 4300 tons, each of the 2074 blocks being placed by hand under the personal direction of the foreman, James Kavanagh of Wicklow.

The base diameter of the tower is 52 ft tapering to 40 ft at a height of 20 ft above the base. Partial rings in this section form the facing to the natural rock face. The total height of the masonry tower is 146 ft 4 in., the external profile being a segment of an ellipse up to a height of 116 ft, and vertical thereafter. Including the lantern room, the overall height of the lighthouse is 179 ft 6 in., the light being 157 ft 11 in. above high water level. The operation of the light became automatic in 1989.

SCOTT C. W. *History of the Fastnet lighthouses*. Hazell, Watson and Viney, London, 1906.

LONG B. *Bright light, white water*. New Island Books, Dublin, 1993, 89–106.

3. Limerick and Killaloe Navigation

To provide a connection between the Shannon Navigation and the sea at Limerick, the Directors General of Inland Navigation commenced work in 1757 on a navigation from Killaloe at the southern end of Lough Derg to the mouth of the Abbey river at Limerick. An artificial cut was made to avoid the Curragour Falls on the River Shannon where it passes through Limerick city between the Thomond and Sarsfield bridges. Ten years later, the work was taken over by the Limerick Navigation Com-

HEW 3078
R 583 574 to
R 705 730

R.C. COX

Derelict lock on the Limerick and Killaloe Navigation

pany, but it was left to the Commissioners to complete the navigation between 1780 and 1785.

The original navigation led downriver from Killaloe, where there was a short bypass canal, to a point east of Cloonlara, where a canal, in places in a 70 ft deep cutting, proceeded to near Cloonlara and then across Newtown Bog to join the Clare Blackwater. This small river was diverted into the canal, which rejoined the Shannon opposite Plassey. Downstream at Reboge a straight cut was made through to Limerick to complete the navigation. There were a total of eleven locks, originally designed to be 74 ft by 14 ft, but actually built to typical dimensions of 120 ft by 24 ft. Design was by Thomas Omer (1757) and later by William Chapman (1767) and the work was carried out by direct labour.

Following the commissioning of the hydroelectric works at Ardnacrusha in 1930, the navigation was allowed to become derelict. A new canal was built from Parteen Villa, where a large weir with floodgates controls the flows in the river. This 7 mile long canal serves as the head-race for the power station at Ardnacrusha. Here there is a double lock, one chamber 40 ft, the other 60 ft deep, to lower vessels of maximum length 105 ft and 19 ft

6 in. beam to the tail-race, which rejoins the main river about one mile downstream.

DELANY R. *Ireland's inland waterways*. Appletree Press, Belfast, 1992, 49–51, 54–58.

4. Limerick Docks

The early development of Limerick as a port included the building of quays on the west bank of the Abbey river and Arthur's Quay and Honan's Quay on the south bank of the River Shannon. At the time of building Wellesley (now Sarsfield) Bridge in the 1820s, the intention was to form a wet dock upstream of the bridge with access provided by a lie-by and locks. Instead the quays were extended westwards along the south bank of the river, but no storm protection was provided for shipping. These quays were well built in local ashlar limestone. Several eminent engineers were consulted and different schemes put forward, including a proposal by Thomas Rhodes for a barrage across the river with locks, but eventually it was decided to construct a wet dock at the western extremity of the then line of quays.

HEW 3011 and HEW 3012
R 567 567

The dock, as originally built, was 800 ft long by an average width of 390 ft, with a 70 ft wide tidal entrance lock at the north-east corner. The work was at first carried out by a contractor named Lawrenson, but was completed in 1853 under the direction of John Long of the Board of Works using direct labour. The lock gates were supplied by J. and R. Mallet of Dublin, but were replaced in 1888 with steel cellular gates by Kincaid & Co. of Greenock. The original dock enclosed about 7½ acres and provided around 2300 ft of wharfage.

A graving dock, 428 ft long overall by 397 ft over the blocks, was constructed in 1873 at the eastern end of the dock with an entrance width of 45 ft (half its length is now filled in to act as a storage compound). The limestone masonry work is particularly fine, with some of the sections of dock being cut directly into the limestone bed-rock. The original wrought-iron floating caisson gate by George Robinson & Co. of Cork is still extant, although badly corroded. Power for dewatering the graving dock was originally provided by a steam engine, the chimney of which can still be seen. There is an impressive clock tower, which contained the hydraulic equipment associated with the operation of the graving dock.

The river wall was rebuilt in 1888 by James Henry & Sons of Belfast under the direction of William J. Hall. The wet dock was extended westwards in 1932, thereby increasing the area to around 11 acres and the length of wharfage to around 3000 ft. A new, more convenient entrance was provided in 1950 at the north-west corner of the docks, the original gates being removed and the north-east entrance filled in. A number of storage facilities were provided, notably Bannatyne Mills on the south side of the docks.

O'SULLIVAN T. F. Limerick harbour dock extension. *Trans. Instn Civ. Engrs Ir.*, 1938, **64**, 179–218.

5. Tralee and Blennerville Ship Canal

HEW 3062
Q 803 137 to
Q 830 140

Plans for the construction of a ship canal to give work to the poor and to improve access to Tralee from the sea were first drawn up by the government engineer Sir Richard Griffith in 1828; an Act was obtained the following year. The design was originally for an impounded canal supplied from springs, but was later altered to provide a tidal canal with an increased depth and width. A number of local contractors were employed between 1832 and 1835 (Leahy and Ferris, 1832; Leahy and Foley, 1832–33; Burke and Foley, 1833–35; and M. and D. Leahy, 1840–46) and progress was slow. The Board of Works then took over the works, and the greater part of the design can be attributed to its engineers, Buck and Owen. The canal finally opened to shipping on 15 April 1846.

The main waterway of the canal is about 74 ft wide at high tide and about 15 ft deep and was intended to accommodate sailing ships of 200–300 tons displacement. The bed and sides are paved throughout to a roughly parabolic section with local limestone slabs. The canal is slightly under 2 miles long and commences at a 30 ft wide sea lock (now derelict) at Annagh on the north side of the estuary of the River Lee. A local road is carried over the canal by a fixed low-level bridge which replaced the original swing bridge. There are lie-bys at this point and beside the sea lock. The terminus of the canal was at a dock or basin 400 ft by 150 ft by 15 ft deep, which was filled in following the abandonment of the navigation in 1945. The dock and locks contain much fine cut limestone masonry, especially around the canal gates, drawbridge and basin, and it is hoped partially to restore the canal as far as Blennerville.

6. Ballyvaughan Water Supply

Ballyvaughan, on the north coast of County Clare, developed as a small fishing village in the nineteenth century.

HEW 3218
M 250 071

Public fountain, Ballyvaughan

R. C. COX

241

It lies on the edge of an area of exposed limestone pavement known as the Burren. Reliable water supplies can be a problem because of the nature of the surrounding geology, much of the rainfall finding its way to numerous underground streams.

In 1872, Lord Annaly built an artificial reservoir on a hillside to the south-east of the village to serve the farming community in the valley. The reservoir, with a capacity of 350 000 gallons, is concave in shape, 84 ft in diameter at top water level and 52 ft diameter at the base. The base and sides are lined with puddled clay to make the reservoir watertight and they rest on the limestone gravel and glacial drift. The reservoir is fed by two springs and the water is distributed through a system of pipes in the normal manner.

The scheme was extended by the Board of Guardians to the centre of Ballyvaughan under the Public Health (Ireland) Act of 1874, the water being conveyed by gravity in cast-iron pipes laid in crevices in the rock or above ground. The engineer for the extended scheme was James Andrews and the contractor was R. Simpson of Dublin. A fountain was erected in the village in 1875 and a number of plaques on it record the names of those involved in providing Ballyvaughan with one of the earliest reliable public water supplies in the country.

ANDREWS J. On Burren waterworks. *Trans. Instn Civ. Engrs Ir.*, 1875, **11**, 12–23.

7. Cork Waterworks

HEW 3077
W 660 718

The main water supply for the city of Cork is abstracted from the River Lee upstream of a weir at Lee Road, about two miles west of the city centre. Following treatment, the water is pumped to a series of reservoirs on the hills to the north of the city and is then fed to the mains. Since 1980, the City Scheme has been augmented from a major multi-million pound Harbour and City Scheme, which takes water from the River Lee at Inishcarra and the River Bandon at Innishannon.

In 1768, the Cork Pipewater Company (established in 1761) installed the first waterwheel and pumps at the Lee Road works to the plans of Davis Ducart. Cork Corpora-

C. RYNNE

tion took over responsibility for the works in 1856 and the City Engineer, Sir John Benson, designed a major scheme of improvements, including the installation of a number of reciprocating pumps driven by water turbines operated by the river flow. The original turbines were installed in 1863 on the site of the present turbines. Three new turbines were installed in 1890 and two more in 1895. Four of the five turbines were supplied by an American company, Jas. Leffel of Ohio. Between 1912 and 1920, new turbine pumps were designed by the waterworks engineer, Michael Riordan, the castings being prepared by the Cork firm of Robert Merrick.

Lee Road Waterworks, Cork

The turbines, and their associated reciprocating pumps, are housed in a fine turbine house. They are preserved, along with some interesting emergency standby steam plant, installed by Combe Barber & Co. of Belfast between 1905 and 1907, consisting of three triple-expansion engines. The pumps at the Lee Road plant are now operated by electricity.

Until 1875, untreated river water was used to supply the city. In that year, a filter tunnel was laid down in the gravel beds located parallel to the river bank. It was designed as an infiltration gallery and consists of a 42 in. diameter brick tunnel some 1500 ft in length. The tunnel terminates in a circular pure water basin from which the filtered water is pumped to the reservoirs. The initial output from the tunnel was reputed to be 5 million gallons per day, but the present output is only around one million gallons per day, as a result of the gradual clogging of the gravel beds. Rapid gravity filters were added in 1928, when chlorination of the water was carried out for the first time.

GRAY G. Cork waterworks – history and development. *Irish Engineers*, 1977, April.

FITZGERALD W.A. *Cork waterworks 1768–1984: history and development.* Cork Corporation, Cork, 1984.

8. Lismore Bridges

HEW 3076
X 047 990

The first bridge at Lismore in County Waterford was commissioned by the 5th Duke of Devonshire of nearby Lismore Castle to carry the Tallowbridge to Cappoquin road over the River Blackwater. It was built between 1774 and 1779 by Darley and Stokes to the design of the architect Thomas Ivory and consisted of a single span of 100 ft over the main channel of the river and an approach viaduct of eight smaller spans (land arches) crossing the flood plain on the north side of the river.

Following severe flood damage in November 1853, it became necessary to replace most of the viaduct, the main arch and adjacent span on the north side being retained. A new approach viaduct, consisting of five segmental arches, was constructed by a contractor called Williams, but some structural weaknesses in this section caused it to collapse again on 26 December 1885, the day before it was to be reopened to the public.

The viaduct was completely rebuilt in 1858 by C. H. Hunt and E. P. Nagle to the design of Charles Tarrant, County Surveyor of Waterford. The present six approach arches range in span from 40 ft 9 in. to 47 ft 3 in. with rises of between 0.30 and 0.35.

These six arches are themselves approached on its north-east side by a three-span masonry arch bridge, also built to Ivory's design. The bridges are all constructed in the local limestone and the approach viaduct arches in particular are of some architectural merit.

R. C. COX

When it was built, the arch spanning the main river channel at Lismore was the largest in Ireland. One unusual feature is that the springing point of the arch at the north pier is 7 ft 3 in. below that of the south abutment because of the profile of the river valley at Lismore. The rise to span ratio of the main arch is about 0.22.

Main span of
Lismore Bridge

9. Causeway Bridge, Dungarvan

Since 1987, Shandon Bridge has carried the main Waterford to Cork road over the River Colligan at Dungarvan. The original line of the road crossed the mouth of the river by the Causeway Bridge. The name derives from a short causeway leading from an area known as Abbeyside, on the opposite bank of the river to the town.

HEW 3061
X 262 934

The building of the Causeway Bridge formed part of the improvements to Dungarvan undertaken at the beginning of the nineteenth century by the 5th and 6th Dukes of Devonshire, whose seat is at Lismore Castle. It is generally accepted that the final design was the work of William Atkinson, the architect to the 6th Duke.

The bridge spans 77 ft 6 in., the segmental arch having a rise of 14 ft to the scroll keystone at the top of the arch. The width between the solid parapets is only 27 ft 8 in.

R. C. COX

Causeway Bridge, Dungarvan

and there is no footpath of any significance. Although suitable limestone was available locally, Atkinson chose to import sandstone from Runcorn in Cheshire. The joints in the spandrel masonry are at right angles to the intrados of the arch and are continuations of the voussoir joints, thus creating a pleasing fan-shaped appearance when viewed from the harbour road. There are similarities with the masonry work found in the Ringsend Bridge at Dublin which was completed in 1812 and is of similar span. The heavy ornamented abutments (to counteract the lateral forces) feature rectangular panels and the masonry blocks were given a rough or 'rusticated' finish.

The work was supervised between 1809 and 1816 by the Yorkshire civil engineer Jesse Hartley, who married a local girl and went on to achieve fame as the engineer responsible for developing the Port of Liverpool between 1824 and 1860. His son John Bernard Hartley, who assisted his father in Liverpool, and carried out important dock works on his own, was born in Dungarvan.

O'KEEFFE P. J. and SIMINGTON T. A. *Irish stone bridges: history and heritage.* Irish Academic Press, Dublin, 1991, 280–281.

10. Thomond Bridge

HEW 3059
R 577 579

The present Thomond Bridge spanning the River Shannon in Limerick city replaced an earlier masonry arch bridge of 14 spans reputedly erected in 1210 in the reign of King John.

The bed of the river at this point is formed by a rock ledge, which provides a firm foundation for the masonry piers. The designer, James Pain, used the foundations of alternate piers in the old bridge as the foundations for the new piers. Erected between 1838 and 1840 by the Board of Works, Thomond Bridge consists of five segmental arches of 50 ft span and 10 ft rise, and two flanking segmental arches of 40 ft span and similar rise. There is a generous width of 36 ft between the solid parapet walls. Pointed cutwaters are used to direct the flow of the river between the piers. Good quality limestone is to be found locally and this material was used by Pain in the Thomond Bridge and in his bridge erected some ten years previously further upstream at Athlunkard.

11. Sarsfield Bridge

Alexander Nimmo, a gifted pupil of the renowned civil engineer Thomas Telford, worked in Ireland between 1825 and 1832, during which time he designed and supervised the construction of many roads, bridges, and piers, mostly in the west of the country. One of the best examples of his work is Wellesley (now Sarsfield) Bridge over the River Shannon in Limerick city.

Nimmo's design for Wellesley Bridge was greatly in-

HEW 3058
R 574 573

Sarsfield Bridge

R. C. COX

Alexander
Nimmo

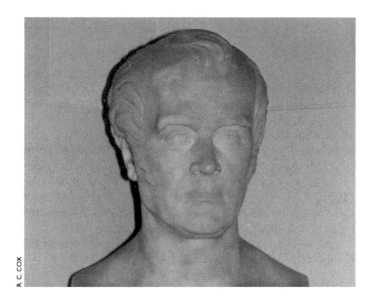

R. C. COX

fluenced by the world famous Pont Neuilly, designed by Jean Perronet, head of the Corps des Ponts et Chaussés, and built across the River Seine in Paris in the 1780s. Ruddock regards Nimmo's design as 'economical as well as scientific and elegant'.

The Wellesley Bridge Act was passed in 1823 and provided the sum of £60 000 for the bridge and some work on the nearby docks. The foundation stone was laid on 25 October 1824, the contractor being Clements & Son. Towards the end of the construction period the works were taken over by the Board of Works and in 1831 John Grantham replaced Nimmo.

The main bridge between the river island and the west bank comprises five arches of limestone ashlar masonry each of 70 ft span and 8 ft 6 in. rise, a rise to span ratio of only 0.12. The height above the waterline at which each of the main arches springs reduces from the face of the arch to the bridge centre line. This was done to aid the hydraulic flow at times when high tide levels coincided with the river being in flood, a condition now controlled to a large extent by the requirements of the hydroelectric power station upstream at Ardnacrusha. The tops of the cutwaters are carved to resemble seashells.

The bridge cut off access from the sea to the existing

harbour and quays. Provision was made for shipping to reach the upper city quays and the Shannon Navigation through a lie-by and entrance channel, spanned by a swivel bridge. The original twin-span swivel bridge supplied by Forrester & Co. of Liverpool was replaced in 1923 by a single leaf swivel bridge from the Cleveland Bridge and Engineering Co. of Darlington. The opening mechanism of this bridge was electrified in 1926 and the bridge fixed permanently in a closed position in 1963.

Wellesley (Sarsfield) Bridge was opened on 5 August 1835 and was declared toll-free in 1883 as a result of liberal grants from the Grand Juries of both Limerick and Clare Counties. The bridge and all the quays upstream are now the responsibility of Limerick Corporation.

RUDDOCK T. *Arch bridges and their builders 1735–1835*. Cambridge University Press, Cambridge, 1979, 198–200.

O'KEEFFE P. J.and SIMINGTON T. A. *Irish stone bridges: history and heritage*. Irish Academic Press, Dublin, 1991, 274.

12. Bealaclugga Bridge

During the 1820s, the government of the day, then based in London, carried out extensive public works in Ireland, including roads and harbours, much of the activity being along the western seaboard.

HEW 3181
R 038 772

Bealaclugga Bridge

R. C. COX

In County Clare, the coast road between Kilkee and Milltown Malbay crosses the Annagh River near Spanish Point. At this location, John Killaly erected a masonry arch bridge in 1824, in what might be called the 'castellated gothic' style, at the same time exercising a considerable flair for ornamentation.

Reminiscent of Alexander Nimmo's bridge over the River Liffey at Pollaphuca, County Wicklow, Bealaclugga Bridge has a span of 34 ft, the rise to the soffit of the pointed arch being 21 ft 9 in. The substantial abutments are over 17 ft thick and 35 ft wide and are heavily ornamented with a mixture of shields, rings and slits.

13. Spectacle Bridge

HEW 3144
R 123 978

In 1875, the County Surveyor of Clare, John Hill, was faced with the problem of carrying the main road from Ennistymon to Lisdoonvarna over the River Aille, where it flows in a narrow gorge, with near vertical sides, in places up to 80 ft high. At the site chosen, the road level is some 46 ft above river level.

Hill provided a novel design aimed at avoiding excessively deep spandrel walls, thereby reducing significantly the dead weight of the filling. He built an ordinary semicircular masonry arch of 18 ft 2 in span at low level. Resting on this he constructed masonry brickwork, with a large circular opening of 18 ft 2 in. diameter, which carries the roadway. The visual effect from river level gave rise to the name Spectacle Bridge.

The overall length of the bridge is 72 ft; the abutments extend 20 ft 4 in. either side of the arch with a width of 24 ft 9 in. and a further 6 ft 3 in. either side with a width of 26 ft 9 in.

Because this unusual bridge is on a major route, and has a carriageway width of only 18 ft with narrow verges, it is vulnerable to being replaced or bypassed. It is to be hoped that County Clare's 120 year-old oddity will be preserved and survive intact.

Spectacle Bridge

R. C. COX

14. Listowel Bridge

In 1822 the government sent their engineer Richard Griffith to the south-west of the country to open up communications with the more remote districts with the aim of
providing access for the garrisons based in Listowel and
other towns on the periphery of the area. Many miles of
roads and several masonry bridges of note were completed between 1823 and 1839. Labour was plentiful and
was employed to provide relief from the distress caused

HEW 3064
Q 996 332

251

by a series of famines, the result of successive potato harvests being affected by bad weather. Potatoes were at the time the staple diet of the rural population throughout most of Ireland.

Listowel Bridge in County Kerry carries an approach road across the River Feale to the south of the town of Listowel. Construction was authorized and financed by Act Geo.IV c.81 and completed in 1829 by local masons, directed by Griffith's agent Hill Clements.

There are five segmental arches, each of 53 ft span and 11 ft 9 in. rise. The simple, but robust, design was carried out in fine quality ashlar limestone, relieved by a archivolt, projecting string course and parapet cap stones. Rectangular section pilasters are carried upwards to the parapets from the tops of the pointed cutwaters. There is a 320 ft long approach ramp on the west side of the river. The width between parapets is only 26 ft 6 in., allowing for a 22 ft carriageway and one footpath.

Other extant Griffith bridges include Feale Bridge, a semi-elliptical arch of 70 ft span, and Gouldbourne Bridge, of 55 ft span, on the Newcastle–Castleisland road.

O'KEEFFE P. J and SIMINGTON T. A. *Irish stone bridges: history and heritage.* Irish Academic Press, Dublin, 1991, 285.

15. Athlunkard Bridge

HEW 3131
R 589 590

The main road north from Limerick City crosses the River Shannon at Athlunkard. Here, in June 1826, began the erection of a substantial limestone masonry bridge for the benefit of the citizens and under the direction of the Directors General of Inland Navigation by the powers given to them by the Irish Parliament just before the Act of Union. The bridge was built in the massive style to withstand the full force of the river in flood, although river levels are now controlled at Parteen Weir, at the start of the intake canal to the hydroelectric works at Ardnacrusha. A plaque on the bridge records the designers as the firm of James and Geo. Richd. Pain (Architects). The Pains were pupils of the famous London architect, John Nash.

The five segmental arches are each of 67 ft span, with a rise of 13 ft 6 in. The abutments are 30 ft thick and the piers each 10 ft 6 in. thick. The width between the solid stone parapets is 26 ft, but this widens to just under 50 ft at a point some 23 ft from the face of the abutments. The bridge is built at right angles to the river and is approached by embankments, 111 ft long on the east side and 235 ft on the west. On the west or Limerick side there is an attractive toll-house, which is now a private residence. Athlunkard Bridge was finally opened in December 1830, having taken four and a half years to complete.

16. Fermoy Bridge

The bridge at Fermoy in County Cork is a fine example of a mid-nineteenth-century masonry structure built to last. The earliest bridge across the River Blackwater at Fermoy was built around 1625 and had many small arches, which tended to impede the flow of the river.

HEW 3070
W 812 985

The present bridge, in ashlar limestone with rough dressed facing, represents the partial rebuilding of an earlier structure. It carries the main Dublin to Cork road over the river on the northern approach to the town. The bridge was completed in 1865 by Joshua Hargreave to the design of A. Oliver Lyons.

There are seven segmental arches, which vary in span from 38 ft 6 in. to 48 ft 4 in. giving a rise to span ratio of

Fermoy Bridge

R. C. COX

around 0.22. The width between the solid parapets is a generous 41 ft, allowing adequate room for modern traffic volumes. Retaining walls contain a 190 ft approach embankment at the northern end and there are rounded cutwaters both upstream and downstream. The bridge is partially sited over a river weir and mill-race, which supplied water to a saw mill and corn mill (both now disused).

The town of Fermoy was on one of the main mail coach routes and provided a base for the Italian immigrant Charles Bianconi, who ran a network of fast coach services to Dublin and elsewhere.

BRUNCARDI N. *The bridge at Fermoy.* Éigse Books, Fermoy, 1985.

17. St Patrick's Bridge, Cork

HEW 3149
W 676 721

The tidal channel of the River Lee in Cork city is spanned between St Patrick's Quay and Merchants Quay by a fine masonry arch bridge of ashlar limestone. An earlier bridge on the site, erected in 1789, was swept away by a great flood in 1853.

The present bridge was erected between 1859 and 1861 and reconstructed in its original form in 1981. It has three arches, having an elliptical or three-centred profile, the arch rings being chamfered to assist water flow. The central span is 62 ft 4 in, and the two flanking spans each 56 ft 9 in., with an arch rise to span ratio of around 0.24. The bridge is nearly 60 ft wide between the parapets, which consist of pierced balustrades. When completed, it was the second widest bridge in either Britain or Ireland, only Westminster Bridge in London being wider. The foundations were taken down to 14 ft below low water ordinary spring tides and the piers were formed from cast-iron caissons filled with concrete, the interior being reinforced with transverse bars. The contractor for the bridge was Joshua Hargreave, the grandson of the man who erected the original bridge.

St Patrick's Bridge was designed by the City Engineer, Sir John Benson, who was both a civil engineer and a distinguished architect, and responsible for a number of public buildings and bridges in the city and county of Cork. He designed a small skewed masonry arch bridge at Carrigrohane on the western approach to the city. This is the only known example of a ribbed masonry bridge in Ireland, a technique which had been in use elsewhere since medieval times to reduce the dead weight of the spandrel fill.

18. Daly's Bridge, Cork

Pedestrian access to public facilities on the south bank of the north channel of the River Lee at Sunday's Well in Cork City is provided by a striking suspension bridge of about 150 ft span. Cork Corporation received the bridge in 1927 as a gift from James Daly of Dalymount in Cork.

HEW 3143
W 657 716

The 4 ft 8 in. wide walkway of timber decking is carried on timbers spanning between iron girders, which in turn are supported by vertical suspenders from the suspension cables. Mass concrete piers of 11 ft square section support the twin towers, 12 ft 6 in. high and 3 ft 4 in. by 1 ft 6 in. in plan. Each tower is formed of angle iron with double-latticed cross bracing, the towers themselves being cross braced at right angles to the bridge centre line. The bridge deck is suspended by vertical hangers from two pairs of 1½ in. diameter twisted steel cables taken

Daly's Bridge, Cork

R. C. COX

over the tops of the towers to anchorages on each bank. The bridge deck rises slightly towards mid-span, the sag in the cables being approx. 6 ft 6 in. below the level of the tops of the supporting towers. The latticed hand railing, comprising 5 ft by 3 ft 10 in. panels, is stayed by brackets carried from the ends of the deck support girders.

Daly's Bridge was designed and built by the London firm of David Rowell, Engineers, to the specification of the Cork City Engineer, Stephen W. Farrington.

19. Cork Rail Tunnel

HEW 3114
W 678 734 to
W 684 722

In its heyday, the Great Southern and Western Railway Company (GS&WR) was the largest in Ireland, with over 1100 miles of track, of which 240 miles were double. It provided rail facilities between Dublin and the south and south-west of the country.

The company was incorporated by Act of Parliament in August, 1844, and John Macneill was appointed Engineer. Mallow was reached from Dublin by March, 1849 and William Dargan was entrusted with a single large contract to complete the line to the Munster capital.

The north Cork countryside is quite hilly and presented much more of a challenge to the railway engineers than had the preceding sections through Kildare, Laois and Tipperary. There are, in addition, a number of significant barriers to the progress of the line, one of which was the River Blackwater immediately to the south of Mallow. The river was spanned here by a ten arch masonry viaduct, which was blown up during the Civil War and subsequently rebuilt to a different design. The terrain near Kilnap and Monard demanded even higher masonry viaducts and considerable rock cuttings.

The high ground to the north of the River Lee in Cork city posed a major challenge for the contractor and the line had to be terminated in October 1849 at a temporary station (Victoria) in the district of the city known as Blackpool. Macneill decided on a tunnel, to be constructed on a downgrade of 1 in 60, and 1355 yd long, from Blackpool to a terminus at river level in Cork at Penrose Quay. Dargan commenced work in 1853 and the tunnel was opened to traffic on 3 December 1855. The present terminus was completed in 1893 near the entrance to the tunnel at Lower Glanmire Road (now called Kent station).

The rail tunnel, the longest in Ireland still in use for rail traffic, was driven in old red sandstone and brick lined throughout, except for a short section in the centre. The tunnel cross-section is about 28 ft wide by 24 ft high, except where it widens to nearly double this width to allow the tracks to diverge into the terminal station.

MURRAY K. A. and McNEILL D. B. *The Great Southern and Western Railway.* Irish Railway Record Society, Dublin, 1976.

20. Cork and Cobh Railway Viaducts

From 1859, Cobh (previously Queenstown) became an important port of call for transatlantic liners (served by tender to an anchorage in Cork Harbour). It was from Cobh in 1838 that the steamship *Sirius* sailed to complete the first Atlantic crossing entirely under steam power.

HEW 3111
W 783 719
(Slatty)
W 781 710
(Belvelly)

Situated on Great Island, Cobh was connected by rail to Cork via a branch leaving the main line at Cobh Junction (now Glounthaune). The branch was opened in 1862, the contractor being Joseph Philip Ronayne of Cork.

To reach the island, the line has to cross both the Slatty and Belvelly channels by means of viaducts. The existing iron structures were designed by Kennett Bayley in the 1880s to replace the original timber viaducts. The larger of the two, the Slatty Viaduct over the channel between Harper's Island and Fota Island, has six 70 ft spans, each comprising a set of three wrought-iron parabolic trusses, the maximum height of the top chord being 11 ft. These trusses are carried on 5 ft diameter cast-iron cylindrical piers infilled with concrete. The abutments are pairs of masonry towers founded on timber piles, and are spaced 17 ft apart at base level and connected by relieving arches of 13 ft 6 in. span and 3 ft rise.

Belvelly Viaduct, which was repaired after being seriously damaged in 1922 during the Civil War, has two 60 ft spans and one of 70 ft, and is of the same general construction.

ROWLEDGE J. W. P. *A regional history of railways, Volume 16: Ireland.* Atlantic Transport Publishers, Penryn, Cornwall, 1995, 130–132.

R.C. COX

Chetwynd Viaduct

21. Chetwynd Viaduct

HEW 3073
W 636 677

To the south-west of Cork lies the important market town of Bandon, once a centre for cotton and woollen manufacture. The Cork and Bandon Railway (later to become the Cork, Bandon and South Coast Railway) was completed in 1849 as far as Ballinhassig. This western part of the line was built to the design of Edmund Leahy by four different contractors (Bolton, Henright, Jones & Parrett, and Condor) with a short tunnel of 170 yd at Kilpatrick, the first in Ireland through which trains passed.

The final 4 miles of the route into Cork city included the construction of the Chetwynd Viaduct with four spans of 110 ft and a headroom of 83 ft, and a tunnel of 900 yd at Gogginshill, just east of Ballinhassig. The viaduct was designed by Charles Nixon, a pupil of Isambard Kingdom Brunel, constructed by Fox Henderson & Company, and opened in 1851. The general contractor was John Philip Ronayne.

The viaduct comprises cast-iron arches, each having four ribs with transverse diagonal bracing and latticed spandrels. The arches span between substantial masonry piers, 20 ft by 30 ft in section. Cross girders originally carried iron plates which supported the single-track permanent way. The structure was repaired following damage sustained during the Civil War in 1922. The line was closed to traffic in 1961 and the decking removed to discourage access. The Chetwynd Viaduct crosses the main Cork to Bandon road and is a listed structure.

ROWLEDGE J. W. P. *A regional history of railways, Volume 16: Ireland.* Atlantic Transport Publishers, Penryn, Cornwall, 1995, 135.

22. Laune Viaduct

In 1881, the Great Southern and Western Railway Company (GS&WR), using powers dating from 1871, decided to construct a branch from the Tralee main line at Farranfore for a distance of 12½ miles to Killorglin. Forming part of the eventual route to Valentia, the branch line ran over fairly level ground, but had to cross the River Laune by a viaduct between steep banks near Killorglin.

HEW 3085
V 780 968

Opened for traffic in 1885, the viaduct has three 95 ft spans, each consisting of pairs of twin cross braced bowstring girders of part N truss and part latticed members. The piers and abutments are of local limestone. Designed by the engineering section of the GS&WR, the viaduct was built by the contractors T. K. Falkiner and S. G. Frazer.

Some corroded members were replaced in 1950. The

Laune Viaduct

R. C. COX

railway closed in 1960 and, following the renewal of the decking and the provision of handrails, the viaduct was reopened in late 1993 as a public footpath to form part of a riverside walk.

ROWLEDGE J. W. P. *A regional history of railways, Volume 16: Ireland.* Atlantic Transport Publishers, Penryn, Cornwall, 1995, 149.

23. Gleensk Viaduct

HEW 3071
V 580 885

The Great Southern and Western Railway Company, having opened up the kingdom of Kerry in the 1850s by their route from Mallow to Killarney and Tralee had, by 1885, completed a branch line from Farranfore Junction to Killorglin. The government of the day, considering the possible use of Valentia as a point of departure for the Atlantic mails, financed, by means of grants and Baronial guarantees, the continuation of the railway at standard gauge down the southern shore of Dingle Bay. At Valentia a ferry operated to Valentia Island, where there was a telegraph and meteorological station.

Opened to Valentia on the 12 September 1893, this 16 mile section of the line included some of the steepest gradients and most spectacular engineering works on any Irish railway. Following a 3 mile climb of 1 in 59 to Mountain Stage, the permanent way was set on a narrow shelf cut into the hillside with the county road and sea far below. There were a number of tunnels and heavy earthworks, including large retaining walls, and a curving viaduct at Gleensk. The line was closed to traffic in February 1960.

The viaduct, designed by A. D. Price and built by T. K. Falkiner, crosses over the valley of the Gleensk river between Glenbeigh and Cahirciveen at a height of 70 ft and is on a 650 ft radius curve, the sharpest on any standard gauge track in the country. The viaduct has eleven spans of 58 ft which comprise steel girders carried on slender masonry piers.

Because of the tight radius, the wheel flanges of the rolling stock of passing trains ran along the sides of the rail. This generated considerable heat and water troughs were provided to cool the flanges.

McGRATH W. Branch lines in County Kerry. *Railway Magazine*, 1956, **102**, 288–294.

24. Ballydehob Viaduct

West of Skibbereen in West Cork lies a remote area leading down to Mizen Head. To serve part of this area, the West Carbury Tramways and Light Railways Company, with the aid of government grants and Baronial guarantees, built a 3 ft gauge line from the terminus of the Cork, Bandon and South Coast Railway at Skibbereen as far the fishing village of Schull. Construction began in 1884 and the line opened to traffic in July 1886.

The only major civil engineering work on the line was the twelve-arch viaduct which carried the single-track line over an inlet of Roaringwater Bay near Ballydehob. The semicircular arches, with 24 in. deep arch rings of rough-hewn coursed limestone, span between mass concrete piers with coursed rubble facings. The piers are 4 ft 3 in. by 14 ft at the base, tapering to 12 ft 6 in. at the springings.

The 18 in. wide parapet walls were raised in recent years to a height of 3 ft 9 in. for safety reasons, the 9 ft 4 in. wide viaduct now forming a pedestrian walkway linking the two sides of the inlet. Circular walks have

HEW 3106
V 990 355

Ballydehob
Viaduct

R. C. COX

been made possible by providing access across a weir built downstream of the viaduct in order to maintain the water level in the estuary.

The designers of the viaduct were S. A. Kirkby and John William Dorman, and the contractors were McKeon, Robinson and d'Avigdor of London. The line was closed in 1947.

ROWLEDGE J. W. P. *A regional history of railways, volume 16: Ireland.* Atlantic Transport Publishers, Penryn, Cornwall, 1995, 140.

NEWHAM A. T. The West Carbury tramways (the Schull and Skibbereen Light Railway), *Journ. Ir. Rlwy Rec. Soc.*, 1964, **6**, 73–97.

25. Mahon Viaduct

HEW 3105
S 393 062

The rail route from Waterford via Dungarvan and Lismore to Fermoy in County Cork was opened in 1878 with the completion of the curving viaduct over the River Mahon near Kilmacthomas. The part of the route between Dungarvan and Fermoy was closed to traffic in the 1960s, but a bulk freight service was operated between Waterford and Ballinacourty up until recently.

The viaduct consists of eight semicircular masonry arches, each spanning 32 ft 6 in. between high piers. Designed by the engineer to the railway company, James Otway, the viaduct was built by Smith Finlayson & Co. of Glasgow. A public recreation area based around an old mill extends along the river and under the viaduct.

26. Carrickabrack Viaduct

HEW 3115
W 825 991

The Fermoy and Lismore Railway was carried over the River Blackwater east of Fermoy in County Cork on a viaduct consisting of a main river span and five side spans on the northern side of the river. This line became known as the 'Duke's Line', as it was almost totally funded by the Duke of Devonshire of Lismore Castle as a means of improving transport to his estates and for the use of the local populace.

The 15¾ miles of railway, authorized in 1869 and built by John Bagnell in 1872 under the direction of James Otway, was taken over by the Waterford, Dungarvan and Limerick Railway in 1893. This company in turn was absorbed by the Fishguard and Rosslare Railway and Harbours Company, who were responsible between 1903 and 1906 for renewing the viaduct. Only the piers and abutments date from 1872. The central span of 157 ft consists of twin 15 ft deep wrought-iron latticed girders with 24 in. wide flanges.

The five approach spans on the Fermoy side of the river are reversed parabolic steel plate girders with rolled I-section cross girders, the average span being 49 ft.

27. Cahir Viaduct

Cahir Viaduct is on the rail route from Limerick to Rosslare Harbour via Waterford, the line having been built in stages between 1848 and 1906. The section from Tipperary to Clonmel was constructed for the Waterford and Limerick Railway (W&LR) between 1851 and 1853 by contractor William Dargan. In order to cross the River Suir at Cahir, the Chief Engineer to the Waterford and Limerick Railway, William Le Fanu, specified wrought-iron box girders, spanning between 16 ft thick limestone masonry abutments and two 10 ft thick limestone masonry river piers. The clear spans are 52 ft, 150 ft and 52 ft

HEW 3108
S 050 251

Cahir Viaduct

R. C. COX

263

respectively, the girders being joined together so as to act continuously over the piers. The box girders, made up of plates and angle-irons, were supplied by William Fairbairn of Manchester, the holder of a patent for such girder bridges.

The box girders are 11 ft 7 in. deep overall (a span/depth ratio of 13 to 1) and 2 ft 6 in. internal width (3 ft wide across the flanges). To provide extra resistance to the compressive forces, the top flange is in the form of a box, 1 ft 3 in. deep and 2 ft 6 in. wide, running the length of the main girders. The boxes are divided internally into cells by vertical plates at 2 ft intervals and are spaced 26 ft 6 in. apart. The permanent way is supported on 9 in. wide by 1 ft 6 in. deep transverse beams spanning between the main box girders. The last 18 of the transverse beams at the Waterford end of the bridge were replaced in mild-steel in 1956, following accidental damage caused by the crash of a loaded sugar beet special train. The viaduct was designed for double line working, but the line was singled in 1931 for reasons of economy. The running track is now laid on the southern side of the viaduct.

The abutment towers have castellated features of some architectural merit. The intermediate piers, as originally constructed, also had these castellated features, but these were removed some years ago.

This fine example of an early box girder railway bridge of large span is in daily use for both passenger and freight traffic. It is thought to be the only essentially unaltered Fairbairn patent type railway bridge still in existence in either Britain or Ireland.

FAIRBAIRN W. On tubular girder bridges. *Min. Proc. Instn Civ. Engrs*, 1849–50, **9**, 223–287.

TYSON S. Notes on the history, development and use of tubes in the construction of bridges. *Industrial Archaeology Review*, 1978, 143–153.

28. Barrington Bridge

HEW 3060
R 680 549

One of the few surviving early iron bridges in Ireland, Barrington Bridge carries the Annacotty to Cappamore road over the Slievenohera river south-east of Limerick city.

The bridge was erected in 1818 by J. Doyle of Limerick

R. C. COX

Barrington Bridge

for Sir Matthew Barrington to gain access from his residence to the parts of his estate situated on the far side of the river. The cast-iron bridge spans 53 ft 9 in. between masonry abutments with a rise of 3 ft 1 in.

The outer ribs are integral with the spandrels and the arches so formed carry a repeated four-leaf clover ornamentation. Immediately above the intrados of the arch there is a pierced design of alternate diamond and oval shapes. Each of the inner ribs consist of an assembly of nine 6 ft long 6 in. diameter cast-iron pipes connected longitudinally by gasket-type flanges at the ends of each pipe. The resulting ribs are tied transversely to the spandrels by ties located above the pipes.

The width is only 13 ft 6 in. and this fact, coupled with a 16 ton weight limit, is restrictive to traffic. There is always the threat that the bridge would be replaced if the route was to be upgraded, in which case it would be preferable to dismantle and re-erect it in another location, for example a public recreation area.

TIERNEY M. *Murroe and Boher: the history of an Irish country parish.* Browne and Nolan, Dublin, 1966.

29. Ballyduff Upper Bridge

HEW 3065
W 964 994

In the 1850s, a timber bridge was erected across the River Blackwater near the village of Ballyduff Upper in County Waterford some 6 miles west of Lismore to connect the Fermoy–Cappoquin and the Fermoy–Lismore main roads. This bridge was replaced in 1887 by the present structure, which consists of twin wrought-iron double latticed girders spanning from abutments of local dressed limestone either side of the river to a central pier, with spans of 99 ft and 103 ft respectively. The girders are joined together in order to act continuously over the pier. The ironwork was supplied and erected by the firm of E. C. and J. Keay of Birmingham, the design being to the requirements of the County Surveyor, W. E. L'Estrange Duffin.

The main girders are 6 ft 9 in. deep, giving a depth to span ratio of approximately 1 to 15. The bridge deck is carried on cross girders, the overall width of the carriageway being only 20 ft. The river pier is 6 ft wide with a pointed cutwater upstream. The northern approach to the bridge is on an embankment contained within flanking retaining walls, which include a small flood arch.

Ballyduff Upper Bridge

Iron bridges were never commonplace in Ireland – timber or masonry were the preferred materials. The

WATERFORD COUNTY COUNCIL

bridge at Ballyduff Upper thus represents an important element of Ireland's extant civil engineering heritage.

30. Eastern Bridges, Kinsale

About 1½ miles to the east of the fishing and yachting town of Kinsale, the main road to Cork crosses over two arms of a tidal estuary, one leading to the village of Brownmills, the other being the Belgooly river. In common with many other locations, timber bridges were constructed. By 1876, these needed to be replaced, and the County Surveyor, S. A. Kirkby, decided on a pair of iron bridges of fairly light construction, but sufficient to accommodate the traffic of the day. The general contractor for the bridges was J. O. and C. E. Brettell of Worcester, and the ironwork was supplied locally by John Steel & Sons of Cork.

HEW 3178
W 658 518 and
W 662 522

The two three-span bridges, the Ringnanean Bridge over the Belgooly river and Eastern Bridge to the south, are of similar construction but differ in span, the former having centre-to-centre spans of 30 ft, the latter 33 ft.

The bridges comprise 24 in. deep girders resting on masonry abutments and on river supports at their one-third points; the supporting columns consist of groups of cast-iron screwed piles bolted together and cross braced. The lateral spacing between the columns is 6 ft, giving a bridge width of approx. 18 ft. The roadway is carried on buckled plates spanning between the main girders. The balustrading is of more recent origin.

The iron bridges have now been bypassed by a new road alignment, but are being retained by Cork County Council for pedestrian use.

31. Mizen Head Bridge

In order to gain access to the lighthouse station on Mizen Head on the south-west coast of County Cork, a footbridge was constructed in 1909 for the Commissioners of Irish Lights across an inlet to connect the mainland with Cloghan Island. A number of designs were considered, including suspension bridges and a steel girder bridge supported on a high central latticed tower. The design

HEW 3026
V 734 234

R.C. COX

Mizen Head
Bridge

chosen, dated December 1908, consists of a reinforced concrete through-arch type bridge. The bridge spans 172 ft at a height of 150 ft above mean sea level.

The main ribs have a parabolic curved profile and a trough section and were cast on the hillside before being slung into position by means of an aerial ropeway. The load-bearing trestles and beams were also cast on shore. The bracing was fixed and the ribs filled in, followed by the fitting of the trestles and beams. The floor and hangers were shuttered and cast *in situ*. The rise of the arch ribs is 30 ft and the ribs are spaced at 15 ft centres at the springing, reducing to 5 ft 6 in. spacing at mid-span. The footway is 4 ft 6 in. wide.

When built, the Mizen Head Bridge was the longest span reinforced concrete arch bridge in either Britain or Ireland and the first of its type. The designer was M. Noel Ridley of Westminster, London, who used the Ridley and Cammell System (an indented steel bar of reinforcement), and the contractors were Alfred Thorne & Son, also of Westminster. Repairs were carried out to the bridge in 1973. It is one of the oldest surviving reinforced concrete structures in Ireland, and probably the first to use precast elements.

In 1997, a visitor centre was opened at the lighthouse

station. The bridge now provides public access to the centre, which displays early photographs of the bridge and drawings of a number of the alternative designs.

ANON. Footbridge at Mizen Head, Ireland. *Conc. & Constr. Eng.*, 1910, **5**, 847–850; 1956, **51**, 30.

32. Kenmare Bridge

At Kenmare in County Kerry, in 1840, as part of a system of government roads which were being constructed in the south-west of the country, William Bald designed a wrought-iron suspension bridge with two half catenaries of pairs of chains. These were suspended from a central tower built on an island in the middle of the estuary of the Roughty river south of the town. This bridge was replaced in 1932–33 by the present reinforced concrete structure designed by Mouchel & Partners of London. It carries the road from Glengariff to Killarney and comprises two 150 ft span parabolic arches passing through the deck.

HEW 3089
V 910 700

The arch ribs rise 34 ft 4 in., are 2 ft 6 in. thick, and 3 ft deep at the crown, deepening to 4 ft 9 in. at the springings. The horizontal deck, about 31 ft wide, is supported by 14 transverse beams, 2 ft 6 in. deep by 1 ft 3 in. wide by 29 ft

Kenmare Bridge

R. C. COX

9 in. long. Ten are suspended from the ribs by hangers at 11 ft 3 in. centres, two rest on short articulated columns near the springings, and two are supported by pairs of columns standing on the rib skew backs at the abutment and central pier respectively.

In 1988–89, there was some replacement of damaged concrete and refurbishment of the reinforcement. The bridge, originally hinged at both ends, is now fixed at one end and hinged at the other.

Opened on 25 March 1933, the Kenmare Bridge was built by A. E. Farr of London.

BUCKLEY C. J. Kenmare Bridge. *Trans. Instn Civ. Engrs Ir.*, 1934, **60**, 29–69.

33. Lee Road Tunnel

HEW 3255
W 725 722

An immersed tube road tunnel is nearing completion (1997) for Cork Corporation. The tunnel passes under the River Lee downstream of the city centre between Mahon and Dunkettle and forms part of the South Ring road system around the city.

The length of the crossing is 610 m and there are two dual carriageways; the northbound and southbound lanes are in separate tunnels with a service tunnel in between. The five reinforced concrete tunnel units are each 122 m long, 24 m wide and 9 m high, and each weighs 27 000 tonnes. The units were constructed on the south bank of the river at Mahon and floated out and submerged into previously dredged trenches.

The Lee Road Tunnel is the only immersed tube tunnel to be constructed in Ireland. It is similar in many respects to the latest Conwy crossing on the north Wales coast.

The design and build main contractor was Tarmac Walls Joint Venture; Ewbank Preece O'hEocha, in association with Symonds Travers Morgan, were retained as the consultants.

34. Shannon Power Scheme (Ardnacrusha)

HEW 3007
R 586 618

As Ireland was emerging from the Civil War in the autumn of 1922, a young physicist and electrical engineer, Thomas McLoughlin, joined the firm of Siemens

Schukert in Berlin, where he devoted a good deal of his
time to developing the concept of harnessing the power
of the River Shannon to produce electricity. He was in
time able to persuade the government of the Irish Free
State to accept a plan to construct a single hydroelectric
power station at Ardnacrusha near Limerick to achieve
the most efficient utilization of the fall in level between
Lough Derg and the Shannon estuary. Equally important
to the success of the Shannon Scheme, as it became
known, was to be an electricity grid, stretching the length
and breadth of the country. Rural electrification was to
become the essential framework for the social, economic
and industrial development of the country.

The Shannon Scheme, commenced in 1925, involved
the construction of a weir and intake near Parteen Villa,

Shannon Power
Scheme
(Ardnacrusha)

ELECTRICITY SUPPLY BOARD

head- and tail-race canals spanned by four bridges, and the power station complex itself. Both the weir and the intake canal are interlinked: the weir across the river regulates the flow of the Shannon and diverts water into the 7½ mile long head-race canal, while the flow in the canal is controlled by the intake. The weir was designed to raise the water level in the river at Parteen Villa to that of Lough Derg, thus ensuring that the entire fall between Killaloe and Limerick could be used to drive the turbines in the power station downstream at Ardnacrusha.

Essentially, the power station consists of an intake sluice house, penstocks, generating building, waste channel and navigation locks. The intake sluice house is built on top of a 405 ft long mass-concrete gravity dam across the end of the head-race canal. This regulates the flow of water through the 131 ft long by 19 ft 8 in. diameter penstock tubes, inclined at a slope of 31° to the vertical shaft Francis and Kaplan type turbines located in the power house. This is a steel-framed building, the foundations of which lie 98 ft below the intake house. Water from the turbines is discharged via 59 ft long draft tubes to the tail-race and is then conveyed back to the river downstream at Parteen-a-Lux. There is a 600 ft long fish pass and navigation locks capable of taking vessels up to 105 ft long.

The Shannon Scheme, one of the largest civil engineering projects of its type in the world at the time it was built, was officially opened on 22 July 1929. By 1935, Ardnacrusha was supplying around 80 per cent of the country's electricity requirements. The turbines and generators have recently been replaced and the capacity of the station increased from its original 82 MW to 110 MW.

WINCHESTER C. (ed.) *Wonders of world engineering*. Fleetway House, London, 1936, 357–367.

DUFFY P. Ardnacrusha—birthplace of the ESB. *North Munster Antiquarian Journal*, 1987, **29**, 68–92.

35. Inishcarra Dam

HEW 3248
W 544 722

The first major harnessing of the power of the River Lee for the generation of electricity began in the 1950s with the construction of dams at Inishcarra and Carrigadrohid, upstream from Cork city.

The dam at Inishcarra is the only example in Ireland of the buttress type. This design effected a considerable saving in concrete compared to the more normal gravity type dam. It was also claimed that the design reduced the possible uplift on the base of the dam, and permitted a flatter slope to be used for the upstream face, thereby adding a substantial weight of water to the stabilizing forces acting on the dam.

The dam is 800 ft long and 140 ft high and divided into 19 blocks, mostly 45 ft long, with a water stop between each block. These massive buttressed blocks have 'diamond headed' upstream faces. The buttresses are 18 ft wide with a downstream slope of 8 : 10. The upstream lead of each block is of constant section with a slope of 3 : 10 and a contact face 6 ft 6 in. wide between blocks.

Upstream at Carrigadrohid, there is a smaller gravity-type dam, which together with the Inishcarra Dam created two roughly equal storage reservoirs with a combined storage capacity of about 5 per cent of the annual flow of the river. The dams were constructed by the French civil engineering firm Société de Construction des Batignolles.

MURPHY A. M. River Lee hydroelectric scheme, Part 1: Civil engineering works. *The Engineers Journal*, 1953, **6**, 498–502.

36. Blennerville Windmill

In 1784, to relieve distress amongst the poor, the government introduced corn laws to encourage farmers to return to tillage by granting bounties on the export of grain. To meet a local demand for meal and flour, the local landowner, Sir Rowland Blennerhasset, erected a five-storey tower windmill at Blennerville near Tralee.

HEW 3063
Q 813 130

The combination of the opening in 1846 of the ship canal from the channel in Tralee Bay to a dock at Tralee, and the repeal of the corn laws in 1848, led to the windmill ceasing to operate and it fell into disuse. The sails were subsequently destroyed by gales and there followed a further period of neglect until 1981 when the Tralee Urban District Council purchased the mill with a view to its restoration. With aid from the state training and employment authority and other agencies, the windmill was

Blennerville
Windmill

TRALEE VISITOR ATTRACTIONS

fully restored and adjacent buildings converted to a heritage centre. The centre focuses, not only on milling, but tells the story of the mass emigration which took place following a series of disastrous famines in Ireland. Many ships left the quayside at Blennerville in the 1840s and 1850s filled with emigrants on assisted passage to the New World and elsewhere.

The overall height of the windmill is 57 ft 8 in. to the top of the conical cap, whose weight, including the windshaft, is around 9.8 tons. The four sails, each 33 ft by 9 ft, and weighing 2.8 tons, are attached to the inclined windshaft. The shaft turns at 10 to 15 revolutions per minute in winds of force 3 to 5.

A platform gallery surrounds the tower at first floor level (9 ft 5 in. above ground level). At an early stage in the windmill's history, it is recorded that Lady Blennerhasset was tragically killed when she stepped out on to this platform and was hit by one of the sail arms.

The base of the tower is 30 ft in diameter with walls 4 ft 2 in. thick, reducing to 26 ft 4 in. diameter at the level of the base of the cap, where the walls are 1 ft 10 in. thick.

Blennerville Windmill is a great tourist attraction, especially since the Tralee–Blennerville section of the Tralee and Dingle Light Railway reopened to passengers in 1992.

KELLY L. *et al. Blennerville : gateway to Tralee's past.* Blennerville Windmill Company, Tralee, 1989, 385–401.

37 Midleton Distillery Waterwheel

In 1825, James Murphy established what to to become the largest whiskey distillery in the Cork harbour area at Midleton. Irish Distillers built a new distillery on a nearby site in 1975 and the original complex of buildings was left virtually intact and is now a working museum. The distillery contains the world's largest copper pot still, with a capacity of 33 333 gallons.

**HEW 3267
W 886 734**

The site also houses one of the largest waterwheels in Ireland still in full working order. The cast-iron wheel, of the suspension type, was supplied in 1852 by William Fairbairn of Manchester. It is 22 ft in diameter and 15 ft wide, with circumferential gearing and ventilated buckets. The water supply is led to the wheel by a masonry aqueduct which predates the present wheel.

McCUTCHEON W. A. *Wheel and spindle: aspects of industrial history.* Blackstaff Press, Belfast, 1977, 16–17.

Midleton
Distillery
Waterwheel

C. RYNNE

Rathmines and Pembroke Main Drainage, **3209**, O 150 320 to O 214 338

Stack C Vaults, **3204**, O 165 347

Thomas Street Windmill, **3104**, O 144 342

Additional Sites

Numbers in bold type indicate Historical Engineering Works (HEW) numbers registered with the Institution of Civil Engineers (ICE) for additional sites within the Republic of Ireland, and for which records are lodged in the archives of both the Institution of Civil Engineers and the Institution of Engineers of Ireland. A number of other sites are under investigation and research is ongoing, leading to the recording of further sites of historical engineering interest.

1. Dublin City and District

Anna Livia Bridge, **3048**, O 102 342

Broadstone Train Shed, **3202**, O 149 352

Chapelizod Bypass, **3220**, O 122 337 to O 087 350

Dublin No. 1 Graving Dock, **3024**, O 184 347

Farmleigh Water/Clock Tower, **3205**, O 091 362

Father Mathew Bridge, **3043**, O 148 342

Great Southern and Western Railway, **3100**, O 136 342

Island (Sarah) Bridge, **3046**, O 128 344

Liffey Rail Bridge, **3054**, O 130 344

Midland Great Western Railway, **3102**, O 148 353

Milltown Viaduct, **3230**, O 166 301

North Bull Bridge, **3093**, O 212 360

O'Donovan Rossa Bridge, **3042**, O 151 341

Phoenix Park Rail Tunnel, **3252**, O 130 343 to O 133 350

2. North Leinster (except Dublin City and District)

Annagassan Bridge, **3029**, O 087 940

Ballinter Bridge, **3171**, N 892 625

Banagher Bridge, **3172**, N 005 159

Bellew Bridge, **3030**, J 018 098

Blundell Aqueduct, **3239**, N 642 313

Boyne Navigation, **3135**, O 090 750 to N 872 678

Brosna Drainage, **3245**, N 340 250 and N 115 245

Dundalk Windmill, **3027**, J 056 075

Gormanston Viaduct, **3057**, O 180 664

Grand Canal (Kilbeggan Branch), **3031**, N 421 256 to N 345 349

Laytown Viaduct, **3034**, O 162 712

Mabes Bridge, **3094**, N 736 768

Marsh's South Sea Embankment, **3028**, J 082 074 to J 072 047

Ryewater Aqueduct, **3129**, N 994 370

Scariff Bridge, **3173**, N 735 526

3. South Leinster (except Dublin City and District)

Alexandra Bridge, **3138**, N 879 270

Athy Church Roof, **3221**, S 683 942

Barrow Rail Bridge (Monasterevan), **3198**, N 623 110

Barrow Rail Bridge (New Ross), **3084**, S 720 294

Bennetts Bridge, **3163**, S 553 493

Bray Harbour, **3052**, O 270 192

Bray Sea Wall and Promenade, **3056**, O 270 188

Brownsbarn Bridge, **3166**, S 618 378

Carragh Bridge, **3141**, N 865 207

Castlecomer Bridge, **3232**, S 538 730

Celbridge Bridge, **3140**, N 973 329

Clara Bridge, **3126**, T 168 920

Dublin Water Supply (Vartry), **3018**, O 218 018 and O 200 270

Dublin Waterworks (Roundwood), **3021**, O 218 018

Goresbridge Bridge, **3162**, S 684 537

Green's Bridge, Kilkenny, **3134**, S 566 506

Liffey Hydroelectric Scheme, **3258**, N 945 085

Mageney Bridge, **3206**, S 718 850

Milltown Bridge (Barrow), **3170**, S 710 505

Rathdrum Viaduct, **3169**, T 193 882

Tacumshin Windmill, **3006**, T 078 075

Taylorstown Viaduct, **3116**, S 820 147

Thomastown Bridge, **3164**, S 586 418

Victoria Bridge, **3119**, N 842 195

4. Belfast City & District

Crawfordsburn Viaduct, **1974**, J 466 817

5. Ulster (except Belfast City & District)

Aranmore Lighthouse, **3213**, B 642 186

Drinnahilly Transit Tower, **1783**, J 361 302

Jerrettspass Canal Bridge, **1900**, J 064 333

Lisnafillan Water Tower, **2101**, D 072 027

Ramelton Water Supply, **3227**, C 217 198

Rathlin O'Birne Lighthouse, **3214**, G 462 797

The Mourne Wall, **1869**, Sq. J 32

Tandragee Feeder Aqueduct, **1901**, J 053 446

Tassagh Viaduct, **1975**, H 861 383

6. Connaught

Athleague Bridge, **3193**, M 827 577

Belmullet Ship Canal, **3196**, F 702 325

Blackrock (Sligo) Lighthouse, **3216**, G 598 400

Eeragh Lighthouse, **3211**, L 758 122

Galway Docks, **3117**, M 300 250

Galway Waterways, **3122**, M 290 250

Inisheer Lighthouse, **3212**, L 976 008

Jamestown Canal, **3194**, N 002 956 to M 980 967

Knockvicar Bridge, **3199**, G 873 055

Lough Allen Canal, **3195**, G 950 054 to 968 110

Mount Talbot Bridge, **3231**, M 812 531

New Bridge, Sligo, **3097**, G 694 360

Newport Bridge, **3200**, L 983 940

Roosky Bridge, **3189**, N 054 860

William O'Brien Bridge, **3121**, M 295 250

7. Munster

Ardfinnan Bridge, **3235**, S 882 176

Ballycotton Lighthouse, **3153**, X 010 635

Ballyvoyle Viaduct, **3145**, S 338 963

Bull Rock Lighthouse, **3152**, V 408 398

Cappoquin Bridge, **3234**, X 100 992

Carrick-on-Suir Bridge, **3168**, S 400 217

Clarke's Bridge, Cork, **3177**, W 671 716

Cork Harbour and Port, **3083**, W 682 719

Feale Bridge, **3184**, R 114 268

Foynes Harbour, **3225**, R 250 518

Galley Head Lighthouse, **3207**, W 340 311

Goulburn Bridge, **3183**, R 170 262

Kanturk Bridge, **3182**, R 383 033

Kilcummer Viaduct, **3233**, R 692 001

Killaloe Bridge, **3174**, R 705 730

Killorglin Bridge, **3087**, V 778 965

Kilnap Viaduct, **3113**, W 665 750

Kilsheelan Bridge, **3236**, S 286 232

Knocklofty Bridge, **3237**, S 143 207

Lispole Viaduct, **3192**, Q 508 009

Mallow Bridge, **3185**, W 561 981

Mallow Viaduct, **3112**, W 540 982

Mine Head Lighthouse, **3156**, X 285 824

O'Brien's Bridge, **3148**, R 665 668

Old Head of Kinsale Lighthouse, **3154**, W 632 393

Poulgorm Bridge, 3263, W 206 354

Sir Thomas's Bridge, **3238**, S 238 228

St Vincent's Bridge, Cork, **3176**, W 667 721

Suir Rail Bridge (Waterford), **3069**, S 582 142

Tarbert Lighthouse, **3215**, R 077 499

Thomas Davis Bridge, Cork, **3175**, W 653 715

Tralee and Dingle Light Railway, **3103**, Q 836 144 to Q 445 009

Valentia Viaduct, **3086**, V 472 800

General Bibliography

Barry M.B. *Across deep waters: Bridges of Ireland*. Frankfort Press, Dublin, 1985.

Casserley H. C. *Outline of Irish railway history*. David and Charles, Newton Abbot, 1974.

Clarke P. *The Royal Canal: the complete story*. Elo Publications, Dublin, 1992.

Conlin S. and De Courcy S. *Anna Liffey: the river of Dublin*. O'Brien Press, Dublin, 1988.

Cox R. C. *Bindon Blood Stoney: a biography of a port engineer*. Institution of Engineers of Ireland, Dublin, 1990.

Delany R. *Ireland's inland waterways*. Appletree Press, Belfast, 1992.

Delany R. *The Grand Canal of Ireland*. David and Charles, Newton Abbot, 1973.

Delany V. T. H. and Delany D. R. *The canals of the south of Ireland*. David and Charles, Newton Abbot, 1966.

Fayle H. *The narrow gauge railways of Ireland*. Greenlake Publications, London, 1946.

Flanagan P. J. *The Ballinamore and Ballyconnell Canal*. David and Charles, Newton Abbot, 1972.

Flanagan P. J. *The Shannon–Erne Waterway*. Wolfhound Press, Dublin, 1994.

Gilligan H. A. *A history of the port of Dublin*. Gill and Macmillan, Dublin, 1988.

Green E. R. R. *Industrial heritage of County Down*. HMSO, Belfast, 1963.

Hague D. B. and Christie, R. *Lighthouses: their architecture, history and archaeology*. Gomer Print, Llandysul, 1975.

Lohan R. *Guide to the archives of the Office of Public Works*. Stationery Office, Dublin, 1994.

Long B. *Bright light, white water*. New Island Books, Dublin, 1993.

McCutcheon W. A. *The canals of the north of Ireland*. David and Charles, Newton Abbot, 1965.

McCutcheon W. A. *The industrial archaeology of Northern Ireland*. HMSO, Belfast, 1980.

Murray K. A. and McNeill D. B. *The Great Southern and Western Railway*. Irish Railway Record Society, Dublin, 1976.

Mulligan F. *150 years of Irish railways*. Appletree Press, Belfast, 1983.

Murray K. A. *The Great Northern Railway (Ireland): past, present and future*. GNR(I), Dublin, 1944.

O'Keeffe P. J and Simington T. A. *Irish stone bridges: history and heritage*. Irish Academic Press, Dublin, 1991.

Patterson E. M. *The Great Northern Railway of Ireland*. Oakwood Press, Lingfield, 1962.

Pearson P. *Dun Laoghaire–Kingstown*. O'Brien Press, Dublin, 1982.

Rennie Sir J. *The theory, formation and construction of British and foreign harbours*. John Weale, London, 1854.

Robb W. *A history of Northern Ireland railways*. Northern Ireland Railways, Belfast, 1982.

Rowledge J. W. P. *A regional history of railways, Volume 16: Ireland*. Atlantic Transport Publishers, Penryn, Cornwall, 1995.

Ruddock T. *Arch bridges and their builders 1785–1835*. Cambridge University Press, Cambridge, 1979.

Scott C. W. *History of the Fastnet lighthouses*. Hazell, Watson and Viney, London, 1906.

Shepherd W. E. *The Dublin and South Eastern Railway*. David and Charles, Newton Abbot, 1974.

Shepherd W. E. *The Midland Great Western Railway of Ireland*. Midland Publishing, Leicester, 1994.

Sweeney C. L. *The rivers of Dublin*. Dublin Corporation, Dublin, 1991.

Sweetnam R. and Nimmons C. *The port of Belfast 1785–1985*. Belfast Harbour Commissioners, Belfast, 1985.

Name Index

Engineers

Architects

Contractors

Subject Index